Astronomers' Universe

Series Editor
Martin Beech, Campion College, The University of Regina
Regina, SK, Canada

The Astronomers' Universe series attracts scientifically curious readers with a passion for astronomy and its related fields. In this series, you will venture beyond the basics to gain a deeper understanding of the cosmos—all from the comfort of your chair.

Our books cover any and all topics related to the scientific study of the Universe and our place in it, exploring discoveries and theories in areas ranging from cosmology and astrophysics to planetary science and astrobiology.

This series bridges the gap between very basic popular science books and higher-level textbooks, providing rigorous, yet digestible forays for the intrepid lay reader. It goes beyond a beginner's level, introducing you to more complex concepts that will expand your knowledge of the cosmos. The books are written in a didactic and descriptive style, including basic mathematics where necessary.

More information about this series at http://www.springer.com/series/6960

Francisco Sánchez

The Rise of Astrophysics in Modern Spain

From Dictatorship to Democracy

Translated and with notes
by T. J. Mahoney

Foreword by
Brian May

 Springer

Francisco Sánchez
Founding Director of the Instituto de Astrofísica de Canarias
La Laguna, Spain

Translated and with notes by T. J. Mahoney
Research Division
Instituto de Astrofísica de Canarias
La Laguna, Spain

English translation of the original Spanish edition published by the Instituto de Astrofísica de Canarias, Spain, 2019, entitled "Soñando estrellas: así nació y se consolidó la astrofísica en España"
ISBN: 978-84-09-13229-4
Depósito legal: TF 652-2019

ISSN 1614-659X   ISSN 2197-6651   (electronic)
Astronomers' Universe
ISBN 978-3-030-66425-1
ISBN 978-3-030-66426-8   (eBook)
https://doi.org/10.1007/978-3-030-66426-8

This Springer imprint is published by the registered company Springer Nature Switzerland AG
The registered company address is: Gewerbestrasse 11, 6330 Cham, Switzerland

The 10.4-m Gran Telescopio Canarias in the final stages of construction. (Credit: IAC.)

To all who have made possible the miracle
of astrophysics in Spain
To my sister of hill and mountain
4m+2s+b

Foreword

The author of this book, Professor Francisco Sánchez, affectionately known as 'Paco' to those of us who are fortunate to know him well, is a pioneer, a tireless warrior for science, and the acknowledged father of astrophysical research in Spain.

I first met Professor Sánchez in 1971, as a young undergraduate student. Pursuing my chosen subject of the study of the zodiacal light by spectroscopy, I arrived in Tenerife with a vanload of homemade electronic gear, a custom-built Fabry–Perot spectrometer, and the component parts of an 'assemble-it-yourself' tin hut, which was to be my 'observatory'. I was well out of my depth and in need of assistance.

My PhD supervisor, Professor James Ring, head of Astrophysics at Imperial College London, had set me on a course of investigating the motions of inter-planetary dust, looking at Doppler shifts in the sunlight reflected from it, seen in the night sky as the beautiful phenomenon of the zodiacal light. Although at its best this diffuse band of light in the night sky can actually outshine the Milky Way, it's still seen by only a few humans—even astronomers—because, to see it, you have to be there at the right time, with dark skies and clear horizons, and you have to be close to the tropics. No chance in London, then, where I was based, so Jim Ring packed me off to Tenerife, acting on a kind offer from Professor Sánchez, who was to keep me under his watchful eye while I set up my gear and attempted to make the observations I needed. So, thanks to Paco, I had the back-up of the University of La Laguna, where Professor Sánchez had recently founded his Astrophysics Department. In those days, of course, I would never have dreamed of calling the professor 'Paco'. His office was at the end of a long echoey corridor in the old University building, and I seldom dared to go knocking on his door. He was already the revered head of a world-beating organization—and a humble student does not feel he can often disturb such an eminent figure. In those days, my grasp of the Spanish language was almost non-existent, and Paco was also not fluent in English, so this also hampered our communication. Nevertheless, in our meetings he was always kindly and very understanding of my problems, and gave me everything I needed to do my work, including the assistance of one of the brightest lights of his team, Carlos Sánchez Magro, now sadly deceased.

The actual location of the observation site was beyond my wildest dreams. Far above the busy commercial life of Tenerife, in the breathtaking national park area of Las Cañadas, was a volcanic ridge known as Izaña, with views of the majestic central peak of Tenerife, El Teide, and, on a clear day, the whole beautiful island. At that time the astronomical observatory consisted of just a small building to provide support for the French zodiacal light telescope, a Spanish solar telescope, and the foundations of the first dome for a large telescope—a collaboration between the University of La Laguna and Imperial College. I also remember an old military meteorological observatory on a nearby hill.

And it was here that my humble hut was taking shape. In those days, my fellow student and I slept on site, and living and working up there was a life-changing experience—no phones, no postman, and no food supplies except what you drove up with! It was a peaceful and quite lonely life, broken up only by making the journey down to Santa Cruz to visit the post office. But the compensations were glorious. Suddenly, as a city-born-and-bred boy of 21 years, I was waking up every day above the sea of clouds, to deep blue skies, and air that is always crisp even when the sunlight is hot, and the silence of Nature, along with a view that seeped into my soul, has never left me. The sky on a good night is dark, transparent, and matchless in its splendour. I developed a love for Tenerife, the tropical night sky, and all things Spanish, which is still strong in me today.

But enough of me! The Spanish ZL building housed the instruments with which Francisco Sánchez had made his observations of the zodiacal light, beginning in 1964, and already published groundbreaking papers on the photometry of this phenomenon. In fact, the strong connection between the ZL and Tenerife goes right back to the 1850s, when Scotland's Astronomer Royal, Charles Piazzi Smyth, made his famous journey to 'site test' on Mt Guajara, and Teide itself. He set out to test Sir Isaac Newton's theory that observing from a high-altitude site, above most of the turbulence of the atmosphere, would secure much better quality astronomical 'seeing'. Piazzi was able to confirm the theory and also made the first spectroscopic study of the zodiacal light.

For the following hundred years, things were quiet in Tenerife as regards astronomy, but in 1959 astronomers flocked to Tenerife to observe a total eclipse of the sun. It was soon after then that Professor Francisco Sánchez began the titanic and miraculous task of creating one of the world's leading international astronomical observatories and the Instituto de Astrofísica de Canarias (IAC).

Francisco Sánchez, the great dreamer, was heart and soul devoted to promoting the excellent quality of the skies above the Canarian summits in order to persuade European astronomers to send their most advanced telescopes to the Canaries and to recruit young Spaniards for training as astrophysicists with the help of their foreign colleagues. His skills had to extend not only to scientific research and theory, but also to make of them formidable people with the organizational skills needed to pilot the funding and development of a major scientific installation with successive Spanish governments.

As a result of Francisco's passion and tireless powers of persuasion, Teide's telescopes were operational in the 1960s and, in an inspiring international spirit of collaboration, under his baton, Teide grew in the ensuing 50 years into one of the best-known astrophysics research centres in the world, in terms of science, technology, training of astrophysicists and technologists, and outreach. Today, the Canarian Observatories occupy many hectares, and house solar and nocturnal telescopes owned by more than 60 institutions from more than 20 countries. Thanks to Paco's vision and genius, Spanish and international astrophysics now operates not only in Tenerife but also at Roque de los Muchachos Observatory on the neighbouring island of La Palma, which now boasts, among its many fine instruments, one of the most powerful telescopes in the world, the Gran Telescopio Canarias (GTC), inaugurated in 2009.

This book tells the history of this extraordinary expansion of Spanish astronomy, in detail, from a personal and human perspective, and I am deeply honoured to have been asked to write this foreword.

As final personal note, Paco and I became great friends in my many successive trips to Tenerife, during which I came to appreciate him as an outstandingly human being, always dedicated to his work, yet never too busy to enjoy wonderful food and drink in the perfumed air of Tenerife. And he did me the honour of opening the doors of his University to me one more time in 2006 to help me complete the PhD studies that I had begun 30 years earlier. My PhD thesis, *A Survey of Radial Velocities in the Zodiacal Dust Cloud*, was finally published in 2007.

This book tells, in personal and human, as well as technical, detail, the story of the extraordinary expansion of Spanish astronomy under Professor

Sánchez's direction, setting Spain today in the forefront of astrophysical research worldwide. Paco has always believed in science as a key element in the future of humanity. For him people are what counts, and he never tires of repeating this view. That is why he considers his greatest triumph to be the high level of prestige attained by those who have been trained at the IAC.

The history of modern astronomy in Tenerife *is* indeed the history of Professor Francisco Sánchez himself!

Enjoy this unique chronicle of one of the great success stories in the history of the exploration of the Universe!

London, UK 2019   Brian May

Preface

There is no path unless it first be trodden.
What your path? What is your destination?
The goal that so enthrals you today will change tomorrow.
What matters is the quest, the distant glow that lights your way.
There is no path unless it first be trodden.
When years have passed and they bid you look back,
That furrow you see in the distance—black, red, silver, and gold—
Will be a story made by many as they passed through.
There is no path unless it first be trodden.

What you will find in this book is the story (an enjoyable one, I hope) of the origins and progress of the Instituto de Astrofísica de Canarias (IAC, or *El Astrofísico*, as it is affectionately known in the Islands), and of astrophysics in Spain generally. This is not an astronomy book that breathlessly relates the wonders of the Cosmos, rather it is a commentary on the so far brief history of astrophysics in Spain.

The birth and flourishing of this science in Spain have been so rapid and spectacular, and took place at such a critical time in the country's history that the idea slowly gelled in my mind to write it down. I have the advantage of having been a front-row witness of, and actor in, the events described, and I have striven to explain the how and why of what occurred. The story is there for all to see.

You will learn how the concept of the astronomical sky of the Canarian summits as a public natural resource has been successfully exploited to the benefit of science, technology, and even tourism while also laying a rich cultural and economic deposit in the country.

I have tried to tell the story simply and—hopefully—entertainingly by sprinkling the facts with anecdotes and events involving important and well-known personalities. I have avoided overloading the text with data, references, and documentation. There is no cumbersome academic apparatus of footnotes and textual bibliographical citations, although there are bibliographical notes in the back matter for those readers who wish to delve further into various aspects of the story told here. Neither will you find exhaustive accounts of scientific discoveries and achievements. All that is now amply available in great detail on the internet (web addresses are given where you will find all the facts and figures).

Since the events described are interrelated and overlap, the book is conceived as a mosaic of self-contained chapters that do not follow a strict chronological order. I hope, nevertheless, that the overall collection brings the global picture into sharp relief.

You will see as you read on that scientific research is heavily influenced by the political and economic vicissitudes of the moment—even more so in Spain, where wealthy patrons are few indeed. But you will also come to realize that this weakness is always surmountable through constancy. Not all the stories told here are as epic or literary as those related in the first chapters. Some chapters may even seem heavy going because of their densely packed content. This book has been a labour of love, but I have throughout endeavoured to be rigorous.

La Laguna, Spain 2019   Francisco Sánchez

Acknowledgements

A number of people have made this book possible, and I would like to acknowledge their collaboration publicly.

My closest ally is Maribel Arévalo, my companion, my friend, and my love. She sifted through each chapter as I wrote it, giving her honest and pointed criticisms, while consigning the early versions to the wastepaper bin.

The second filter —literary and grammatical—was the philologist Juncal García-Ramos Arévalo, a cultivated and well-read lady of exquisite taste.

Thirdly, I express my gratitude to Casiana Muñoz-Tuñón, astrophysicist, my ally in many battles, a deep and dependable person. Among other things, she prevented me from occasionally committing overheated and inappropriate remarks to paper.

With her dedicated, meticulous, and whole-hearted approach to all that the IAC represents, Mariam Mónica Gutiérrez, a very special person, must receive particular mention. She undertook the delicate and complex task of repeatedly formatting and correcting the text.

Terry Mahoney, with his usual professionalism, took on the task of translating the book into English. He also aptly proposed a number of additions to the Spanish version.

Dr Carmen del Puerto, our resident journalist and former director of the Science and Cosmos Museum, among her many other duties, and currently head of the IAC's Press and Outreach Unit and defender of women in science, supervised the entire production process of the Spanish version of the book, including its attractive cover, designed by Inés Bonet.

To all, my deepest thanks.

Contents

1   Perhaps It All Began with an Eclipse (1959–1961)   1

2   Izaña (1961–1963)   13

3   Tenerife and Mount Teide on the Moon (1724–1914)   21

4   We Are Stardust (1962–1968)   29

5   The Zodiacal Light (1962–1982)   37

6   Astronomical Site Testing (1961–1974)   45

7   Astrophysics Versus Astronomy (1961–1972)   55

8   Astrophysics Enters the Spanish University
    Curriculum (1970–1985)   61

9   The Reluctant Chancellor (1976–1980)   71

10 A Multinational Astrophysics Treaty (1960–2014)   79

11 Telescopes Versus Military Radars (1977–1978)   97

12 Science Meets Politics: The Law of Astrophysics
    and Its Avatars (1975–2011)   105

13 Royalty and Heads of State Above the Clouds (1985)   119

14 The Sky Law (1978–2017)   131

15 The IAC: A Dream Come True (1971–Present)   137

16 Towards Excellence in Research (1971–Present)   151

17 Astronomical Instrumentation, Technology
    Transfer, and Its Impact on Industry (1974–Present)   161

18 The Rise of Space Astrophysics in Spain (1942–Present)   179

19 Training Future Generations: The Canary Islands
    Winter School of Astrophysics (1989–Present)   189

20 Bringing Astronomy to the Public (1986–Present)   195

21 The Biggest, Most Advanced Telescope
    in the World (1989–Present)   211

22 The European Extremely Large Telescope:
    Dealings with ESO (1961–Present)   229

23 The Starlight Foundation: A Step Beyond
    Astrotourism (2008–Present)   247

24 Astrophysics in Spain: The Wider Picture
    (1974–Present)   259

Epilogue   269

Biblographical Notes   271

Index   283

1

Perhaps It All Began with an Eclipse (1959–1961)

An eclipse of the sun, one of the most awe-inspiring and impressive spectacles of nature, has always surprised, excited, moved, and terrorized humanity. I have witnessed them on several occasions. That the celestial lamp which gives us light, warmth, and life should suddenly vanish, bringing night in the middle of the day, is at once unnerving, dramatic, and mysterious! What if the sudden darkness were to last forever? All periods and cultures have sought advantage from such fearsome events. To be able to 'divine' such a happening was crucial and evident proof of possessing direct and special communication with the gods. Being able to predict an eclipse was a powerful accomplishment that undoubtedly spurred the development of astronomy.

Today's scientists still continue to observe total eclipses, which provide good opportunities to make measurements in very special conditions. The few minutes of totality, with the sun's disc covered by the moon, are used the better to observe the solar corona and other celestial objects. There are dedicated amateur and professional astronomers who chase eclipses. That is why, on the occasion of the 1959 eclipse, which was seen in totality in the Canaries, astronomers and astrophysicists from the world over flocked to the archipelago.

I have taken upon myself the task of writing the history of the spectacular birth and flowering of astrophysics in Spain and following the path of the great advances experienced during the latter half of the twentieth century in our understanding of the Universe. On the assumption that the 1959 total eclipse of the sun in the Canaries might be considered to mark the birth of astrophysics in Spain, I have decided to begin the story with that event. The establishment and subsequent burgeoning of this branch of science in Spain is another story that will gradually unfold in later chapters of the book.

© The Author(s), under exclusive license to Springer Nature Switzerland AG 2021
F. Sánchez, *The Rise of Astrophysics in Modern Spain*, Astronomers' Universe,
https://doi.org/10.1007/978-3-030-66426-8_1

The Canarian press made much of the eclipse, publishing features days before and after the event, complete with photographs of previous eclipses and the arrival of planeloads of scientific equipment. The islanders have not forgotten that extraordinary celestial novelty.

It is useful to recall some of the events that took place back then. The first two lunar probes were launched by the USSR in that year. The second of these, *Lunik 2*, landed on the lunar surface precisely two days before Khrushchev began his visit to the United States. In January, the bearded army of Fidel Castro entered victorious into Havana. In December US President, General Eisenhower, paid a visit to General Franco and was cheered in the streets of Madrid by more than a million Spaniards. Severo Ochoa received the Nobel Prize for Medicine in that same year.

As a consequence of the Second World War, by mid-century scientific and technological advances had produced powerful and precise scientific instruments that greatly increased our observing capacity of both the macro- and microcosmos. Astronomers were now confronted with a situation in which the resolving power of their modern telescopes (their ability to discern two narrowly separated celestial bodies) was limited by the atmosphere in which the telescopes were immersed. It was now quite evident that traditional observatories (often located in national capitals or in the home town of some patron) failed to meet the conditions demanded by modern astronomy (astrophysics). Not even Mount Palomar, where the Americans had just located their 5-metre reflector, the leviathan of its time, could fulfil these requirements.

The new telescopes had to be set up in very special locations with clear, transparent, and stable atmospheres. For this reason, the most developed countries set about searching worldwide for such rare and exclusive sites. Obviously, a good site would have to be cloudless for most of the year and possess a transparent atmosphere. Such conditions prevail in many parts of the planet but are not, of themselves, sufficient. The goal is to identify a site where the upper masses of air are stable enough for celestial objects to be observed as if there were no atmosphere at all, as if the telescope were located in space. Hence, present-day astronomical site prospection is centred on parameters linked to atmospheric turbulence.

This can be understood simply if we explain that the beautiful twinkling of the stars has the same origin as the shimmering of distant objects when viewed above a heated road or behind the flames of a bonfire. Rays of light, which would normally reach our eyes or a telescope mirror in straight parallel lines, are bent when passing through moving bubbles of hot air of a refractive index different from that of cool air.

Fig. 1.1   The local newspaper *El Día* announcing the total eclipse of the sun seen from the archipelago. (Credit: *El Día*)

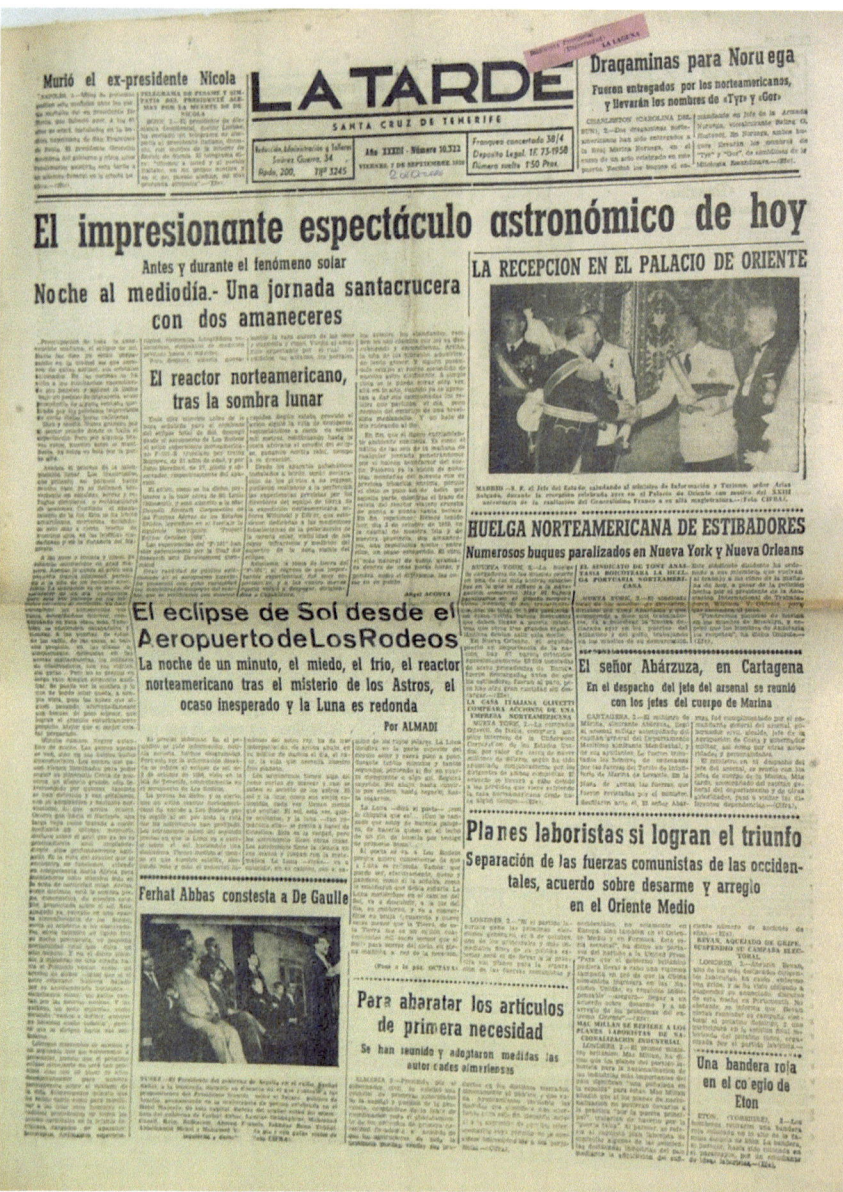

Fig. 1.2 The local evening newspaper *La Tarde* summarizing the eclipse. (Credit: *La Tarde*)

Wavefronts that meet the eye or the mirror of a telescope in this way are no longer plane but corrugated. In sites such as the Canarian summits the stars do not twinkle and their coruscations no longer blur the distant objects beyond them. The performance of telescopes is therefore substantially improved. To further complicate matters, as we increase its diameter, the telescope mirror becomes ever more sensitive to such atmospheric disturbances, and we must be stringent in our requirements concerning the choice of observing site.

The preparations for the 1959 eclipse in the Canarian archipelago prompted many astronomers to recall earlier scientific expeditions that told of the marvels of the sky seen from the island summits. They urged their Spanish colleagues to take steps to ensure that the Canary Islands be included among the sites under consideration for future large telescopes. For this to occur Spain needed to get involved immediately by carrying out astronomical site prospection campaigns on the summits of Tenerife.

The task of bringing this international message to the attention of the government fell to José María Torroja, Professor of Astronomy and Geodesy of the Madrid Complutense University, and Father Antonio Romañá, Director of the Ebro Jesuit Observatory, a major figure in the then recently created Upper Council for Science and Research (CSIC, after its Spanish initials). Both had been trying in vain for three or four years to bring about the creation of an astronomical observatory within the precincts of the Izaña Meteorological Observatory. With the coming of the eclipse, the Spanish authorities (more to save face than anything else) finally allowed themselves to be persuaded, and the Ministry of National Education created—on paper—the Teide Observatory, to be placed under the authority of the Chancellor of the University of La Laguna. Professor Torroja was appointed its director. When it comes to science, our politicians need to be shamed into taking action.

Understanding any story requires that we locate ourselves in the place and time concerned with all the attendant circumstances. While research was forging ahead in more advanced countries, in Spain it languished. University lecturers did little or no research but were nevertheless regarded as demigods (students were obliged to attend examinations wearing a tie, and to stand when lecturers wafted by in the corridor). That was nothing compared to the chancellors, who were exalted beings until, a quarter of a century later, their numbers swelled with the explosive growth of universities in Spain. I mention these details to provide a backdrop to my own entry onto the scene.

I was attracted to research and in my final undergraduate year I was permitted to 'work' at CSIC's Daza de Valdés Optics Institute, a rare privilege at the time for an undergraduate. How I got the 'job' is a curious story in itself. I went to see Dr Plaza, Director of the Institute, and explained to him my wish to do research. He listened to me, asked me questions, and finally told me that I could begin by reporting to him on the following Sunday morning at eight o'clock in the morning. I arrived to be met by him and another researcher, equipped with picks, shovels, and wicker baskets, who turned out to be the head of the Photometry and Colorimetry Laboratory. They explained that it was necessary to demolish a large Ulbricht sphere in order to expand the laboratory, and that the job fell to us owing to shortage of funds, so would I mind lending a hand? It took us five Sunday mornings to complete the task. So when I am asked what my first job in research was, I always say that it involved picks and shovels.

I graduated in Physical Sciences in June, 1960. In September of that year I was managing three precarious jobs: assistant lecturer in the Optics Department of the Complutense University (600 pesetas a month—3.60 euros); supply teacher at the Ramiro de Maeztu Institute (on call in the common room to give any absentee lecturer's class, be it in physics, chemistry, mathematics, or natural sciences); and 'research student' at CSIC's Optics Institute (unpaid). I had a fiancée in Cáceres whom I was unable to marry—not through any lack of ardour on my part (we had been close for many years and I loved her dearly)—but because not even with all these jobs could I maintain a family. As a recent graduate it was the only way to give my career the boost it needed.

In the autumn of that same year I was summoned to a meeting by no less an authority than Professor Navarro, Chancellor of the University of La Laguna, and Professor Torroja. They were thinking of offering me a post in the Canary Islands. It is not hard to imagine the trepidation and nervousness with which I, a recent graduate in Physical Sciences, felt at the thought of being interviewed by two such august beings.

I recall perfectly the room in the CSIC Students' Residence in Madrid where they awaited me. They began by telling me that an astronomical observatory with future possibilities had been set up near Mount Teide under the supervision of the Chancellery of the University of La Laguna. They ended by offering me a contract to carry out 'astronomical prospection' for the site. When I pointed out to them that the term 'astronomical prospection' was completely new to me, they encouraged me by saying that, with my experience in optics and photometry, together with what they knew about me, they had no doubt that I would learn quickly and could indeed carry out such a prospection. They proposed that I spend two years near the Teide, at a site 2400 metres above sea level, taking measurements to determine whether the summits of Tenerife met the criteria for setting up an astronomical observatory. But neither of them could tell me, then or later, the precise nature of astronomical prospection, what observations needed to be made, or how. In answer to my hesitancy and questions they promised to provide me with an unpublished report written by Professor Redman, Director of the Cambridge Observatory, who had just spent a month in Tenerife. All this gave me the misguided impression that the matter should be straightforward.

In the autumn of 1960 two further possibilities of employment presented themselves: a naval post involving magnetic mines, and the setting up and running of a control laboratory for the copper industry. For the first time I became aware that making any choice involved rejecting other opportunities leaving aside a disquieting infinitude of further possibilities.

The choice involving magnetic mines was not difficult to turn down; much more painful was saying no to the economically attractive control laboratory post. Even at that time there was much talk of Spain's joining the Common Market, and the copper sector of our country had opted to prepare for such an eventuality. The industry was looking for a physicist capable of establishing and running a quality control laboratory while also being prepared to take on the engineering establishment in order to achieve 'European quality' for its copper. This was explained to me by Professor Durán, who held the Chair of Optics (I was his assistant) and second-in-command of the Nuclear Energy Council. The offer was: 12,000 pesetas per month (72 euros) on signing of the contract, a year's apprenticeship in the United States and Germany with an honorarium of 24,000 pesetas (144 euros), and, on returning to Spain, that same salary, plus a house and car in Asturias.

The Teide Observatory offer amounted to a two-year contract, 4500 pesetas a month (27 euros), living at 2400 metres above sea level, and being stationed at the Izaña Meteorological Observatory of the Air Force (in Spain meteorology was still under military control at the time). After some haggling, they agreed to pay me 500 pesetas (3 euros) more with the added condition that my future wife assist me in my tasks.

The copper option would clearly have been the sensible choice to make: quite apart from such a high salary, it would have meant foreign travel (every ambitious Spanish graduate's dream) and new horizons. But I feared that, after the first few years, the job would settle down to being an office boss in charge of subalterns to carry out systematic analyses of copper samples, signing certifications, and being at endless loggerheads with stick-in-the-mud production engineers. Not a very enticing prospect. Even worse would have been having to give up my dream of becoming a research worker.

The Canarian option was not at all clear. What exactly had to be done and what future lay in store for an observatory that existed only on paper in the

Official State Bulletin were quite unspecified. But the offer stimulated my imagination and held the promise of adventure. It was also quite exotic. Moreover, I thought it would permit me to keep my ties to the Optics Institute and follow up my incipient optical research.

I discussed all this with my fiancée, María ('Mary') Anselma Almeida, and managed to convince her to join me in the adventure, much to the disappointment of family and close friends, and to the fury of Professor Durán, who for years after felt affronted at my turning down the cushy copper job.

Once the decision had been made, we prepared everything in a month, got married on 29 December, and embarked on our honeymoon voyage to the Canary Islands. Our haste quite alarmed our families and raised eyebrows among our acquaintances.

The wedding was splendid and the guests were amply regaled. After leaving the bridal bouquet at the foot of the Virgen de la Montaña, the patron saint of Cáceres, we changed and left on our honeymoon voyage to the Canaries, having spent our wedding night at the Mérida parador and New Year's Eve in Seville, before finally boarding the *Ernesto Anastasia* in Cádiz.

In early January of 1961 we arrived, tired and seasick, at Tenerife after an uncomfortable crossing from Cádiz. Mary had been so seasick during the voyage that she mistook an abandon-ship drill for the real thing and rejoiced in the thought that they would transfer her to a lifeboat so that she could watch the loathed heaving vessel vanish from sight. When we came on deck on the last day, the ship was sailing steadily without pitching; we caught site of the mountains of Anaga. We were awestruck at so unexpected and beautiful a vision; the island seemed to have suddenly risen from the ocean. A bright dawn daubed the admirable and dramatic landscape of the bay of Santa Cruz with brilliant and limpid colours. The jutting rocks of the gorges of Anaga seemed to want to embrace the hull of the ship, and large fluffy clouds, illumined in the growing light, hung from the sky. We never forgot this glorious first encounter, and the rough sea passage faded from our minds.

We had booked lodgings at the Colegio Mayor San Fernando in an apartment for university teaching staff. We took a taxi to La Cuesta, the old road to La Laguna. We passed the neighbourhoods of Vuelta de los Pájaros and Vuelta de las Flores, and stopped at the Mirador de Vistabella. These poetic names accentuated the impressive view that lay spread beneath us: a reverberant indigo ocean with scattered brilliant white clouds, flowers, unknown to us, of riotous colours all around, poinsettias and bougainvillias, which we had never seen before. And all bathed in a brilliant, almost blinding light. After the wintry mainland, we thought that we had reached Paradise.

On the following day we had a meeting with two people who had been employed to help us, and whose contracts were even more precarious than ours. One was Miguel Zalote, a mathematician and meteorologist, who had been discharged from military service after (quite rightly) calling an air force general a fool; the other was Ignacio Izeta, a young chemist, who hankered after a job at the oil refinery, the only industry on the island and the Mecca of all Canarian chemistry graduates. Having cut our family ties and left our homeland behind, we needed the friendship of our new companions more than we realized. They were an essential support to help us survive in our new world. We talked a lot, asking many questions, and they told us a great deal, but all in such a negative and defeatist tone as to put a damper on all our hopes. We concluded that the least of our worries would be the appalling living conditions in Izaña and the lack of equipment to carry out our tasks; much worse were the non-existent observing programme and the abandonment by those who were meant to supervise us. The coolness of the welcome came as a shock to us. Undaunted, however, we put the best face we could on matters and cheerfully concluded that these tidings were merely the exaggerated claims of people lacking in motivation.

After Epiphany, accompanied by Izeta, we were taken in the Chancellor's car up to the Izaña Meteorological Observatory. We inched our way along the dorsal road, crossing the mountains of Esperanza, bristling with slender Canarian pines and laurel forests of a thousand hues of green. The trees began to thin out, gradually giving way to *retama* brush and shrubs, and in the distance the volcano began to assume prominence with its blackish, ochre, and brown lava deposits resting on multicoloured lapilli strata. On reaching the rim of the Orotava Valley we penetrated the cloud layer and, once above the layer, we saw it! There was Father Teide, its snowy peak sitting in state on a throne of lava! It was far more impressive, majestic, and beautiful than we could have imagined.

We approached the tormented landscape of Las Cañadas del Teide, entering a barren volcanic wasteland that seemed to have been ploughed by a Cyclops with huge piles of reddish-brown basaltic boulders, like caramel-coated almonds, and peppered with obsidian bombs of jet-black brilliance. In spite of their centuries of calm, dry petrified rivers of lava seemed still to flow from the summit to form agitated seas on reaching the Cañadas caldera. Close by we could see monstrous agglomerations of solid lava surrounding the Teide, with its ancient overlapping layers. We were overwhelmed, intoxicated, enraptured, incapable of absorbing such a wealth of beauty all at once. The impact of that first view remains with me even today in spite of having

Fig. 1.3 The road to Izaña Meteorological Observatory. From left to right: Mary Almeida, Francisco Sánchez, Miguel Zalote leaning on Chancellor Navarro's car, with Ignacio Izeta in crouching in the foreground. (Credit: F. Sánchez)

trodden that landscape thousands of times at all hours of the day and night, and in all seasons of the year.

We left the Cañadas caldera late in the afternoon, the low sun throwing into sharp relief the outline of the crests that make up the old volcano. Izaña Observatory welcomed us enveloped in a dense fog that almost totally extinguished the dusk, the outlines of the buildings becoming visible only when we came right up to them. A fierce, icy wind almost prevented us from opening the car doors. Two friendly-looking ruffians (so they appeared to my wife) presented themselves. they had black beards and were covered up to the eyebrows with military capes. They collected our things with friendly grunts and took us to one of the buildings, the 'Kaiser's Lodge', which was to be our dwelling at the Observatory. This was no ironic nickname, but referred to an old and dilapidated hunting lodge that the German Emperor had used in Africa and later donated to a group of German scientists at the turn of the century. As we entered that dried, half rotted wooden cabin it groaned under the buffeting it was receiving from the gale. The cold was glacial. It was already night and small lightbulbs with reddish filaments that gave out a dim light flickered to the throbbing of a single-cylinder motor that could be heard nearby. Everything was in half shadow with dark mysterious corners. The driver had gone, Izeta had left, the Observatory had disappeared, everybody was gone, and we were left alone in our unfamiliar surroundings. In a little while even the electric lights expired. We had to light candles that they had given us for just such an eventuality. Exploring this shadowy world with a small torch merely served to make us yet more apprehensive when we pushed the creaking doors of the few rooms we could enter to discover beds and lumber piled up as if there had been a raid. It had been God knows how long since anybody had lived in that ramshackle cabin.

I shall never forget that night of horrors. Mary and I slept fitfully huddled together, not so much because we were newlyweds but simply to ward off the cold and fear. We could hear creatures, no doubt rodents, scurrying under the floorboards while cold, wet draughts licked our heads. In my brief moments of sleep, I lived and relived a recurring nightmare based on the heaving log cabin in *Gold Rush*, the hapless hero being me instead of Charlie Chaplin.

We quickly became aware that the tools at our disposal for doing astronomical prospection were few and, to make matters even worse, defective and unsuited to the task in hand. On seeing this, the bleak panorama that Izeta and Zalote had painted did not seem to us to be quite so exaggerated after all. Once on site it became abundantly clear to us that nobody knew how to carry

Fig. 1.4 The Kaiser's Lodge in better days when located at Cañada de la Grieta. In the foreground, scientists (doctors, physiologists, and the astronomer Mascart), who came to Tenerife in 1910 to observe the return of Halley's Comet. (Reproduced from Mascart's *Impressions et observations dans un voyage à Tenerife*)

out site prospecting: the tools available were inappropriate, and there seemed to be no genuine interest in remedying this absurd situation on the part of those who had sent us on this wild goose chase.

In the face of this harsh reality my wife and I reached crisis point and were tempted to board the first boat home. Why not? Why should I stubbornly insist on our staying? Why should I set about fixing old instruments and designing new ones to make proper measurements? Why battle with wind and tide against all odds? I suppose it must have been Destiny, that mysterious hand which in Greek tragedy impels humanity to challenge the gods and render the impossible possible. As well blame Providence or Fortune.

2

Izaña (1961–1963)

In daytime under a bright sun Izaña and its occupants took on a quite different aspect and were no longer so frightening. The night of horrors had passed
and the beautiful light of a clear dawn chased away our nightmares. The fog
lifted to reveal a landscape that was impressive, lovely, sharply defined nearby
and limitless in the distance, with the snow-capped Teide above and the sea all
round. How different from the flat, open, and bounded horizons we were
used to contemplating in Extremadura and Castile!

In the absence of mist and nocturnal darkness, the Kaiser's Lodge, our first
dwelling at Izaña Observatory, now looked pretty and bucolic from the outside (if you overlooked its peeling varnish and other signs of abandonment).
It must once have been a splendid colonial wooden bungalow. As a building,
it had its charms in spite of having suffered years of neglect and ill use. It was
first of all a hunting lodge in Germany's African colonies, a gift from Wilhelm
II to one of his sons and heirs. The bungalow was later given to a team of
European scientists who came to Tenerife to observe Comet Halley in 1910.
It was set up in Cañada de la Grieta in Las Cañadas caldera. From there it
found its way to Izaña when Spanish meteorologists began their operations
there. Today it wears a coat of cement and forms the hub of a military outpost
for telecommunications.

Izaña Observatory, located on the summit of a mountain ridge of that
name, comprised a scattered group of large, ugly, poorly maintained buildings. It started life, however, as an 'aereological observatory', a base from
which to sound the troposphere with balloons and kites; it was then a thriving
research centre, well equipped and maintained, in which Spain's first modern
meteorologists were trained. It was the cradle of Spanish meteorology and an

13
F. Sánchez, *The Rise of Astrophysics in Modern Spain*, Astronomers' Universe,
https://doi.org/10.1007/978-3-030-66426-8_2

international point of reference at the beginning of the twentieth century. By the time we arrived there, in the early sixties, it was a mere lumber room of antiquated, abandoned measuring instruments, one example more of the state of decay into which Spanish research centres had fallen after the Civil War. Only a few observing instruments remained active that were essential for meeting the Observatory's obligation to broadcast, using a military transmitter, a handful of synoptic data a few times a day. This duty fell to a 'meteorological observer', who had been trained solely for that task. The observer was employed by the National Meteorological Service, which was part of the Air Ministry, the meteorologists and their assistants themselves being part of the military. Many years were to pass until, at the close of the last century, Izaña Observatory would be reopened for meteorological research. It is now once again an thriving site for this branch of science.

Nevertheless, in the sixties the Observatory had a very large staff: the Chief, a military adjutant, who resided in Puerto de la Cruz; two meteorological observers, who did weekly shifts; Alejandro, a master carpenter who maintained the kites, which had not been used for years; Manuel the maintenance man, who looked after the emergency generator; Maruca, the cook; Isabel, the cleaner; three workmen (Pablo and Juan, who did odd jobs, and Santiago, who lived in Orotava, did errands, and delivered the post). Requisitions were recorded in a black oilskin-covered notebook and brought up by a muleteer— yes, a muleteer, complete with mule and all—who climbed the old routes twice a week. Incidentally, photographs of him, together with those of other muleteers used by Mascart (Chapter 3) have given rise to the myth that I brought telescopes up the mountain by mule. On Mondays a grey lorry from Aviation came chugging up the mountain carrying a driver and two relief soldiers to deliver heavy supplies, such as barrels of oil. Too many idle hands with too little to do.

Seen in retrospect, the people who worked at the Observatory were a rogues' gallery of characters with patinas and profiles worthy of a Buñuel film or Cela novel. Coarse, badly dressed, unoccupied and ill-temperedly milling around those old, dilapidated, damp buildings, they had lived in isolation for many years on the 2400-metre-high mountaintop, and their personalities had been chiselled by that harsh environment and enforced daily contact with the other inmates. To top it all, they had little work to do since their functions, existing only on paper, had been defined when the Observatory had been a busy research centre open to the world. Now that their only mission was to read four instruments, take notes, and radio a synoptic report there was no reason for so many buildings and so many staff.

Fig. 2.1 A muleteer fetches supplies to Izaña Observatory. In the background is the building that housed the first site prospection instruments. (Credit: F. Sánchez)

We spent only a few days at the Kaiser's lodge and managed to acquire accommodation in the best room of the main building in Izaña. It was on the first floor, with two doors onto the stairs and two windows. It had two beds, a small wardrobe, and two bedside tables. There was a communal bathroom outside. It was perishing cold. Apart from there being no heating, the wind entered through gaps in the two windows, swept through the room, and whistled through crevices in the badly fitted doors. Without exaggeration, it seemed to have been designed for the specific purpose of inflicting colds, pharyngitis, bronchitis, and pneumonia.

Since the Observatory Chief spent only a few hours a week in his office, we used it for our living room and workplace because it had a chimney and we could keep ourselves warm. On frequent occasions, the door of the office would suddenly burst open, some inquisitive heads poke in, and a voice boom, 'And those are the astrophysicists'. These interruptions would be the culmination of an improvised tour of Izaña, with one of the simple inhabitants of the Observatory acting as guide, pointing out the four meteorological instruments in action and, most importantly, the transmitter. We were the final attraction, a final exotic touch: the astrophysicists. We were on exhibit like monkeys in a zoo.

Fig. 2.2 The main building of the Izaña Meteorological Observatory, where the author and his family lived during the years of astronomical site prospection. (Credit: F. Sánchez)

While all this was going on, the most powerful (25 GeV) particle accelerator of its day was being built in Switzerland, and birth control pills were being put on sale, bringing about a sexual revolution that has given rise to sociological and cultural consequences that are still being played out today.

As time passed, when we began to have children, we managed to persuade the authorities to let us live in the unused accommodation of the Observatory Chief. It was in such a state of dilapidation that, not only rain and snow, but even hailstones found their way through the cracks in the rickety windows. But at last we had a house to live in with a certain degree of independence.

The electricity supply was a serious problem since the fragile cable from Güímar would sever at the merest hint of a storm. In emergencies, a small DKW generator was used for powering a number of light bulbs, whose filaments would glow a dim red, throbbing in unison with the two-cylinder motor. The deposit would be filled with petrol in late afternoon, and the light would hold out until the fuel was exhausted, usually about three hours later. By that time, we had to have had supper and gone to bed. After that, it was candles and torches only, so we acquired the knack of doing everything, including bottle-feeding the children and changing their nappies, with a torch tucked under our arm to keep our hands free. Frontal torches had not yet been invented.

I cannot resist saying a little about the people we were living among. They were good people, but isolation had caused them to blend affection with hatred and to squabble among themselves amidst the never-ending gossip. Any trivial matter could build up into a storm and bring them to brink of murder, as once happened when an inmate chased after another with an axe. That said, in the face of any misfortune or problem they would close ranks; their solidarity was total, and they would help one another out when needed.

The central figure was Maruca, the cook, a slight woman in her fifties, always dressed in black; she was lean and angular, with deep wrinkles that did not hide how beautiful she must once have been in her youth. Always on the go, with her hair held in a bun, she would have looked like a witch had it not been for her permanent state of good humour, which she exercised on those around her in an ironic, saucy, and amusing way. She offered steaming black coffee, made in the traditional way, to everybody at all hours. She would taunt my young wife every morning, making her blush when we arrived for breakfast with, 'And has the charming young lady had a restful night?'

Maruca was single when she arrived at Izaña Observatory when she was twenty years old with a romance behind her and two daughters in train. She raised her daughters, her granddaughters, and—so I believe—her great granddaughters. She was a strong woman, to whom life had not been kind, but she bore herself with dignity, and her sense of humour never deserted her. She had been so long in Izaña that she had even witnessed the fire that had destroyed the old wooden observatory, reputed to have been beautiful and warm. She knew the stories, warts and all, of all who had passed through the Observatory, including the 'bosses from Madrid', as she called those who directed the nation's meteorology from the Retiro Park. She remembered a thousand incidents, which she wittily related. She told us of a singular individual who would ride a non-existent motorbike at full speed among the retamas, breaking and accelerating, while mimicking the noise of the motor.

When we arrived at Izaña, Maruca's granddaughters Pili and Conchita were living with her. Their mother had just died of tuberculosis, a parting gift from her husband, a handsome, shameless, feckless individual who, according to Maruca, contrived to survive the illness and continue to lead a carefree life after having abandoned his daughters. Maruca's other daughter was married to Juan, one of the workmen, a strong, handsome, pale, red-cheeked man with deep-blue eyes. He was helpful and straightforward, but irascible and prone to violent outbursts.

Then there was Isabel the cleaner, corpulent and bad-tempered, always hugging her lapdog Capricho, whose snout she was beginning to adopt. She was God-fearing, superstitious, easily shocked, and ever scandalized by Maruca's shafts of wit.

Following the advice of the Observatory Chief, we kept a certain distance from all of them to avoid taking sides and involving ourselves in their personal problems. We kept ourselves to ourselves, which made us feel even more isolated in our enforced retreat, although we were always available, and they knew it, to help out. My wife even set up a mini-school to provide a basic education for the children on site. Since we spent various years among them while living in Izaña, sharing in their joys and sorrows, and facing common problems, we got to know them fairly well and grew fond of them. I think the feeling was mutual.

I refrain from commenting further on the individuals of the Izaña tribe, or from entering into detail with further anecdotes. It would certainly be an entertaining description of local custom and anthropology, but would unduly prolong the narrative, my aim here being merely to paint an impression of Izaña as a backdrop against which the embryo of astrophysics in Spain began to take shape.

I add only that these people had a keen nose for when our astronomical project hit a bad patch; that is to say, when funding grew scarce and our bosses' abandonment and lack of interest manifested itself. Service to 'the astronomers' would then drop off: no longer would they bring us wood for the fireplace and stove, or water to the house, so there I would be, going to fetch wood among the *retamas* and draw water from the well. After all, we were on loan to the Observatory, and they were under no obligation to serve us, whatever decisions might have been made in Madrid. How lonely and marooned all this made us feel!

While we were attending to these chores at the surreal Teide Observatory, at other—real—observatories in 1963 they were discovering quasars (quasi-stellar objects) by observing the optical counterparts of these mysterious radio sources and galaxies in the remotest confines of the Universe. They were also discovering giant comets, such as the distant Comet Humason (C/1961 R1).

Contrast that with me wrapping up warm after supper ready to go and observe. Outside a wintry night would await me at several degrees below zero with a wind ready to chill me to the bone. I would begin by putting on a long-sleeved shirt, long johns, and thick woollen socks that I had bought in the Plaza de Pontejos before leaving Madrid (I still keep them as a memento). Over all this, layers and layers of protective clothing. I had to walk half a kilometre to reach the amateur telescope to observe diffraction rings in order to assess the astronomical quality of the sky. I would walk out under a star-studded sky. As I carried on walking, the sky would fill with more and more

Fig. 2.3 Some of the Izaña Observatory staff. (Credit: F. Sánchez)

stars; I could even discern the bright clouds of the Milky Way and see the zodiacal light. The lights of the Observatory would promptly disappear and I would be left in the utmost solitude, lost in the Cosmos. The silhouette of the Teide and the Pole Star fixed me beneath the mighty celestial vault. The silence could be felt, broken only by the rhythmic crunch of the frozen ground beneath my boots.

I remember one night I began to hear breathing behind my back. My senses sharpened. I could not see anybody. I stopped and the breathing stopped. I peered into the darkness and saw that I was alone. I walked on and started to hear the breathing behind me again. I was becoming more spooked by the second. I quickened my pace and broke into a run. I was truly alarmed. Breathless and sweating, I reached the small dome housing the telescope and came to the realization that it was my own breathing that accompanied and followed me. It was my physical self that was manifesting itself so that I should not feel so alone in the immense solitude that enveloped me.

These things were moulding our character, toughening us up, and preparing us for the hard future to come. Untried youngsters that we were, we could not see any future benefit. In truth, it was difficult to keep focused and contemplate our endless miseries without becoming disheartened. Of course, being post-war children, we would not readily give in; we had been taught, and believed, that we should be 'impervious to disappointment'.

In recalling these things, I cannot but feel a deep gratitude and great tenderness for my deceased wife, María Anselma Almeida, the mother of my children. A beauty from a provincial capital brusquely transplanted to remote, strange, and very challenging surroundings. But her upbringing had instilled in her the responsibilities of marriage and motherhood, which were understood in that epoch to be sufficient reason for any sacrifice. Those first years were hard. I keep a black and white photograph in a prominent place at home that vividly expresses the panorama. There it stands like a still from a surrealist film: a beautiful, surprised, and anxious young woman standing before a peeling wall, clutching three boisterous, almost identical children. Without her, my life and work would have been very different.

Fig. 2.4 My wife, Mary, with three of our children standing before our lodgings in Izaña. (Credit: F. Sánchez)

3

Tenerife and Mount Teide on the Moon (1724–1914)

In the Mare Imbrium ('Sea of Showers') on the Moon there is a mountain called Pico (the Teide was known as Pico Teneriffe in former times). A little to the north-west of Pico, also in the Sea of Showers, there is a small mountain range called the Teneriffe Mountains. There is an interesting story behind these names.

Lunar surface features have been given names, but the story of how this nomenclature came into being is not widely known. The first to name lunar features was the Flemish astronomer van Langren (Langrenus), who published his lunar map in 1642 under the patronage of Felipe IV of Spain. He named one of the most prominent craters after his patron, Philippi IV (it was later renamed Copernicus). Langrenus also denominated several lunar features after other Spanish monarchs and nobility.

New nomenclatures continued to be introduced at the whim of their creators until, in 1651, the Italian astronomer Riccioli attempted to introduce order into lunar nomenclature by assigning the names of scientists to lunar features. However, neither was he entirely impartial: he gave his own name and that of his disciple Grimaldi to two conspicuous craters on the Moon's western limb. As he was a great admirer of Tycho Brahe, he named the most prominent crater on the Moon after his idol, while consigning the heliocentrists Kepler, Copernicus, and Galileo to the Ocean of Storms (Oceanus Procellarum). In spite of the foibles of its creator, it is on the basis of Riccioli's system that the modern nomenclature of our satellite has been built, so that only names relating to science are now to be found on the Moon. Today, the International Astronomical Union is the arbiter in all matters relating to astronomical nomenclature.

F. Sánchez, *The Rise of Astrophysics in Modern Spain*, Astronomers' Universe, https://doi.org/10.1007/978-3-030-66426-8_3

The name Pico was introduced by the German astronomer Johann H. Schröter to commemorate Mount Teide (then known as Pico Teneriffe). At the beginning of the nineteenth century, following the practice common at the time of naming lunar orographical features after prominent terrestrial formations, Schröter placed Pico on the Moon. The Teneriffe Mountains were added in the second half of the nineteenth century by William R. Birt in honour of the expedition to Tenerife of his friend Charles Piazzi Smyth. That journey, as we shall see later, was to have a great impact on the future of astronomy in the Canaries.

The eighteenth century was one of great scientific circumnavigations, which were to continue into the following century, and astronomers were always involved. Many of these expeditions stopped over at Tenerife. There are records of astronomical observations carried out on the island by Louis Feuillée of Paris Observatory in 1724, Claret de Fleurieu in 1769, and J. C. Borda, A. G. Pingré, and J. R. Verdun de la Crenne in 1778.

The nineteenth century abounded in scientific, and sometimes not so scientific, expeditions to remote countries, although not all were as celebrated as that of Darwin to the Galapagos Islands, the findings of which were to be so decisive for the theory of the evolution of species. It seems that Darwin, inspired by Humboldt's description of his stay in Tenerife en route to America in the final years of the eighteenth century, was eager to acquire a knowledge of Tenerife. But his hopes were dashed when the *Beagle*, on approaching the Santa Cruz road on 6 January 1832, was denied entry on suspicion that it might be carrying cholera on board. The crew did not wish to delay the voyage by the mandatory twelve-day quarantine period, and Darwin had to be content with viewing Mount Teide from afar. In a letter to his father later that year he wrote:

> We were becalmed for a day between Teneriffe and the Grand Canary, and here I first experienced any enjoyment. The view was glorious. The Peak of Teneriffe was seen amongst the clouds like another world. Our only drawback was the extreme wish of visiting this glorious island.

It was in the mid-nineteenth century when Charles Piazzi Smyth undertook his astronomical expedition, which is considered the origin of the mountain observatories that later sprouted up worldwide, especially during the twentieth century.

It was in the winter of 1855–56, through the offices of the Astronomer Royal, G. B. Airy, that Piazzi Smyth, who had been appointed Astronomer Royal for Scotland in 1846, succeeded in persuading the British Admiralty of the need for an expedition to Tenerife to test, at long last, the long-standing belief of astronomers that the stars could better be seen by locating telescopes as high as possible. As Piazzi Smyth surmised before the Council of the Royal Astronomical Society, 'By how much may astronomical observation be improved by the elimination of the lower third part of the atmosphere?'

Piazzi Smyth based the justification for funding his proposed expedition on a reference from Newton, who, in his *Opticks* (1730), had written:

> Telescopes...cannot be so formed as to take away that confusion of the Rays which arises from the Tremors of the Atmosphere. The only Remedy is a most serene and quiet Air, such as may perhaps be found on the tops of the highest Mountains above the grosser Clouds.

The first budget presented by the Admiralty (at the time astronomy was a strategic science for navigation and was controlled by the Admiralty) amounted to 300 pounds, a sum that was immediately raised to 500 pounds, but that still fell short. Piazzi Smyth, however, was determined not to exceed his initial estimate and resolved the difficulty by seeking the shortfall from private sources. With the help of these patrons, he was finally able to honeymoon (he had just married Jessica Duncan) on the Tenerife voyage. The newlyweds spent thirteen nights together on the island in 1856, most of them spent observing above the clouds. When they were not above the clouds, they resided in what is today the Hotel Marquesa, which continues to receive tourists in the centre of Puerto de la Cruz.

The aim of the expedition was to prove that astronomical observations improved notably with increasing altitude. Piazzi Smyth and his wife performed observations from sea level to just below the very peak of the Teide. They set up base first on the ridge that marks the summit of Guajara, the second-highest mountain on the island, at 2717 metres above sea level. From the south it overlooks the magnificent Las Cañadas caldera, the enigmatic Ucanca Valley at its feet, so reminiscent of a Martian landscape (now frequented by UFO devotees in their ceremonies). They then crossed the caldera with their instruments to Alta Vista, at an altitude of 3250 metres, close to the summit of Mount Teide.

They observed the Moon, the planets, double stars, and the zodiacal light, all with the specific purpose of determining whether image sharpness improved above the lower troposphere. Above 2000 metres, it was evident that the diffraction rings of the stars were indeed clear and well-defined, and that close double stars could be separated to a degree impossible from Edinburgh. They also measured ultraviolet radiation from the sun and infrared radiation from the moon, for which reason Smyth is considered to be one of the pioneers of infrared astronomy. They also carried out geophysical and meteorological observations.

All this served to reinforce the notion that making astronomical observations from high mountain sites indeed had clear advantages and improved telescopic performance. Piazzi Smyth's expedition was the precursor and origin of the mountain observatories that began to be built from that time onwards.

So satisfied was he with his results and the facilities provided by the Spanish authorities ('always predisposed on that island to further the aims of the scientists from any country') that he intended to return in later years.

Piazzi Smyth presented his results in a brief report to the British government in October of that same year; he then gave a more detailed summary on 2 June 1857 before the Royal Astronomical Society. In the following year, he published his famous book *Teneriffe, an Astronomer's Experiment of 1856*, which bore the subtitle *Specialities of a Residence above the Clouds*. This lavish publication contained twenty stereoscopic photographic albumen prints, taken by Jessica Duncan Piazzi Smyth. It was the first book to include stereoscopic photographs and is greatly prized among collectors. I am the proud owner of a copy, given to me in 1985 by the Royal Greenwich Observatory. Apart from containing the scientific results of the expedition, it is also a travel journal, in which the author gives his impressions of his journey through an exotic land. He described landscapes, peoples, and customs that would have caught the attention of the English reader of the mid-nineteenth century.

Piazzi Smyth is also known for his detailed measurements of the Great Pyramid of Giza and its astronomical alignments, for which reason he is a precursor of archaeoastronomy. Unfortunately, he became so obsessed with this pyramid that he came to consider it to be of divine origin, even to the extent that he believed its dimensions to be magic numbers enshrining hidden prophesies. It was constructed, he asserted, by Jews under the supervision of Melchizedek himself. Piazzi Smyth defined the 'divine inch', on the basis of relations between the Pyramid's dimensions and astronomical parameters, to be equal to 1001 Imperial inches, in the vain hope that this would be the template for a universal system of measurements. He was therefore a staunch opponent of the decimal metric system, which he considered to be the work of French atheists. These matters have besmirched his scientific achievements (he has a poor reputation among astronomers and Egyptologists alike), while being eagerly cited by numerologists and esotericists.

The writings of Charles Piazzi Smyth encouraged other astronomers to come to Tenerife. In 1890, O. Simony studied variations in the solar

Fig. 3.1 *Top:* Charles Piazzi Smyth standing behind his telescope on the peak of Guajara, with Mount Teide in the background. *Bottom:* Jessie Duncan, Piazzi Smyth's wife, at the encampment on Guajara. (Reproduced from Piazzi Smyth's *Teneriffe, an Astronomer's Experiment*. Credit: ROE/RSE Archives)

Fig. 3.2 Piazzi Smyth's makeshift observatory at Altavista (now hosting a mountain refuge of that name), just below the peak of Mount Teide. (Credit: ROE/RSE Archives)

spectrum with height. In 1895 and 1896 Knut Ångström measured how the solar constant varied with height (Alta Vista, Las Cañadas, Puerto de la Cruz, and Güímar).

The year 1910 marked the return of Halley's Comet, and many came to the mountaintops of Tenerife to observe it. Among them was Jean M. Mascart of the Paris Observatory, who formed part of a well-organized expedition with medical doctors and physiologists interested in studying the damage that any poisonous gases from the comet's tail might produce as it swept past the Earth. How strange it must have been to see doctors and astronomers keeping a close watch on Halley's Comet. They set up their observatory in Cañada de la Grieta under the ramparts of the circular walls of Las Cañadas, using the hunting lodge that the Kaiser used on his African safaris, now earmarked for scientific use—the very same Kaiser's Lodge, in which my wife and I spent our first night on our Canarian adventure (see Chap. 1) at Izaña Meteorological Observatory.

While the medical researchers sunbathed, using themselves as guinea pigs to determine the effects of height and solar radiation on the human body,

Fig. 3.3 The French astronomer Jean Mascart next to one of his telescopes at Guajara, on the same site used a half a century earlier by Piazzi Smyth. (Reproduced from Mascart's *Impressions et observations dans un voyage à Ténérife*)

Mascart, using mules to carry his telescopes, ascended Guajara and set up camp on the same site used by Piazzi Smyth in the previous century.

Mascart was a celebrity in the Paris of that epoch. He reported his voyage to Tenerife in stages in such newspapers as *Le Figaro*, *Revue Générale des Sciences*, and *Nature*. He also wrote for the local newspapers, including *Nivaria*, *El País*, and *Diario de Tenerife*. He even ended up being portrayed as an eccentric savant in the form of the astronomer Scarmat (an anagram of 'Mascart') in a popular comic of the time, *L'idée fixe du savant Cosinus* (1899). In truth, the photographs in his book *Impressions et observations dans un voyage à Ténérife* do portray him as an Indiana Jones-type explorer.

Our summits so impressed the French and Germans that they campaigned for intergovernmental negotiations to be started for the building of a European observatory on the summit of Guajara, but the First World War put an abrupt end to this European project.

At about that time the German astronomers G. Müller, E. Kron, and K. Schwarzschild were performing astronomical observations. However, they were denied access to the summits on the suspicion that they might use their telescopes to spy on the movement of enemy ships (Spain had stayed neutral in the conflict).

All these expeditions, which may be considered the forerunners of Spanish astrophysics, eventually gathered dust until they were remembered by foreign astrophysicists who came to the Canary Islands for the 1959 total eclipse of the sun described earlier (Chap. 1).

4

We Are Stardust (1962–1968)

To state that we are stardust is not simply poetic licence. I did not know that when I arrived in the Canaries at the beginning of 1961. Astrophysicists have convincingly demonstrated that the elements that make up our bodies and all that surrounds us could not have originated here on Earth, on the other planets, or in the sun that warms us: we are born of interstellar dust. But I did not discover this astonishing fact until I visited the Paris Institute of Astrophysics way back in 1962.

I had come to Tenerife to determine whether the skies above its summits met the demands of modern observational astronomy, but neither I nor anybody else in the country knew how to carry out an astronomical site prospection campaign. Thanks to an exchange grant from the International Astronomical Union, I went to Europe to find out. In January 1962, Dr J. Dommanget, on a similar grant, came to visit us. He was one of a team dedicated to a site-testing campaign then being carried out in South Africa and Chile to choose a site for the future European Southern Observatory (ESO). On this trip I began to familiarize myself with the techniques and procedures used in astronomical site testing while at the same time finding out about astrophysics.

The atmosphere has always been a bugbear to astronomers. To a greater or lesser degree, its perturbing presence persistently impairs the observation of celestial bodies. It is always there before our eyes and telescopes, partially or totally obscuring our vision. It acts as a selective and variable filter always degrading the information reaching us from space, whether in the form of visible light or any other kind of radiation. If the Earth had no atmosphere,

F. Sánchez, *The Rise of Astrophysics in Modern Spain*, Astronomers' Universe, https://doi.org/10.1007/978-3-030-66426-8_4

we would see the stars sharply defined against a black background by day and by night, as occurs when observations are made from space.

The increase in telescope diameters, together with improvements in observing instruments, accentuates to an ever-greater degree the unwanted effects of the atmosphere and necessitates more sophisticated site testing when choosing new astrophysical locations. It was to learn all about this that I went to France in March 1962, three days after the birth Jorge, our first child.

It was my first trip abroad, and I went laden with all the prejudices and complexes of Spaniards of that era. In Hendaye, during the long five-hour wait for the French train (the railways of our neighbouring country took this precaution because of the habitual delays of RENFE, Spain's national rail company), I experienced what it was like to be one of the masses that went to work in Europe, with all the fears and home-sickness felt by migrants (of course, our travails could not be compared with those of today's migrants who risk their lives in precarious dinghies). I was sorely tempted to take the return train and go back to my family. I felt torn by having left my newborn son so soon in order to fulfil obligations that I could not simply dismiss.

I shall never forget the way the gendarmes ill-treated the immigrants, shouting at them and shoving them like cattle. Many years later I witnessed a similar scene on the frontier between Nepal and Tibet: a Chinese police-woman, small and frail, used a whip to keep back a group of tall Tibetans who were numb with cold to prevent them from approaching the barrier. In Hendaye I felt the impotent rage of the immigrant at seeing how a gendarme was abusing a Spaniard. I scolded him and was on the point of being beaten and incarcerated. I was young and remembered the history of the Second of May, 1808, in Madrid; I felt a surge of patriotism and heroism. Later—much later—I would learn to abandon my Hispanic inferiority complex. And now, very much later, I have come to realize the extremes to which Spaniards accepted the calumnies of the Black Legend of our inferiority in comparison with northern Europeans. I have also shaken off other misconceptions. It is now abundantly plain that the heights reached by Spanish astrophysics in so little time demonstrate our intellectual and moral equality.

During one of the first stops made by the train, a gentleman entered our compartment and sat next to me; he was pleasant and began to speak to me in Spanish with a strong Catalan accent. His destination was close to Paris; the train would be stopping at every station so we were able to converse for a long time. We spoke of many things, but one topic made a particularly deep impression on me; it was the clarity and depth of his ideas on Latin America—'Hispanic America', as he termed it. He meant by it an entity encompassing all the Latin American countries that share our language, in the same way that

Fig. 4.1 Spanish migrant workers in the sixties about to board a train. (Reproduced from Félix Santos, *Exiliados y emigrados: 1939–1999*. Credit: www.cervantesvirtual.com)

Spain incorporated Catalonia and other regions. He made me change my way of seeing the world of Latin America and view it more accurate. His analysis of the attitudes of Spanish governments towards the Hispanic nations throughout history was thought-provoking and insightful. He lamented the short-sighted policy of the Spanish politicians of the day, who wallowed in sentimentalism instead of extracting the maximum benefit—including economic—from our shared language and culture. It was vital, he said, to establish genuinely closer ties and share business opportunities.

He told me his name was Josep Tarradellas, and that he was the President-in-exile of the *Generalitat de Catalunya*. I was therefore particularly struck by the national way in which this Catalan spoke of Spain and the issues of the day. Listening to him brought to my mind what they used to tell us in Youth Front camps about 'loving Spain in spite of ourselves' in order to put things right. That is not to say that the Honourable Tarradellas liked everything about us, but he did present ideas and remedies to improve things. The propaganda of the Franco regime painted Tarradellas as an anti-Spanish, evil, and conspiring separatist, a public enemy. He seemed to me on the contrary to be a charming and cultivated person of broad political vision, with a great gift for presenting things in a simple and accessible way. I keep that chance meeting as a special memory. I learned many things from him. We exchanged addresses on parting, and I later began to receive letters from him from time to time, together with his talks and political propaganda. I treasure these with great affection. It is tragic how different, supremacist, and antiscientific is the discourse of today's separatists.

Paris was an experience for me in every sense, and not just because of the revelation that we are stardust. There I was, having been educated under National Catholicism, weighed down with all the prejudices and suppressions of Franco's Spain, suddenly immersed in this magnificent city. My eyes were opened wide to a new world. You can imagine my astonishment and excitement. Couples freely kissed passionately as they waited for traffic lights to change. Nobody looked at them with disapproval, and no policeman intervened to fine them. Most daring films were being shown in the cinemas, and there were amusing caricatures of the President of the Republic—General de Gaulle himself!—that you could buy without the risk of imprisonment. Unbelievable!

I had the good fortune to reside and work at the Paris Astrophysics Institute. Through my dormitory window I could see the domes of the renowned *Observatoire* in which so many astronomical discoveries had been made in past centuries. It was located in the Latin Quarter itself, close to the Sorbonne, the Panthéon, Notre Dame, Sainte Chapelle, the Luxembourg Gardens, etc.—a truly privileged site.

There was an air of *coup d'état* in France at the time. The country had been shaken by the bloody excesses of the OAS (*Organisation de l'Armée Secrète*) and Paris was under police control. I was greatly shocked to see manned machine gun posts set up on sandbagged barricades at the entrances to police stations and a number of public buildings. It was just like a Second World War film. They even stopped me in the street and demanded to see my passport as they prodded the barrel of a rifle against my torso. It was far from amusing, but added a certain zest to my explorations of Paris.

Some years later, during another of my visits to Paris, I was again staying at the Paris Astrophysics Institute and my visit chanced to coincide with the May 1968 uprising. Living as I was in the Latin Quarter, I saw the beginning of the protests with my own eyes: paving stones were being lifted from the streets, and there were barricades and violent confrontations. I was so alarmed that I took the last train back to Spain before they closed the frontier. I cannot, then, presume, like so many politicians of the Transition, to count myself among the 'French May' heroes.

But back to '62. Curiosity has always prompted me to find things out for myself. Having little money and not minding walking, I set about getting to

Fig. 4.2 A protest march for peace in Algeria and against OAS terrorism. (Credit: André Cros, Municipal Archives of Toulouse Town Hall, under licence CC BY-SA 4.0)

know the city on foot. During the afternoons and on feast days I would take the metro to a distant part of town and from there, street guide in hand, I would return on foot, visiting the monuments and streets that I had previously chosen. I noted many things, most of them unremarkable but surprising to me. I visited museums, of course, and admired the buildings and avenues like any other tourist; I blended in with the variety of people that thronged the streets. I enjoyed watching and absorbing all that was happening around me, and enjoyed the cheap meals offered by small restaurants. Until then I had never liked cheese, yet here I was ordering unknown items from the menu that would invariably be cheese or something made of cheese. As my francs were in limited supply and the walking had stimulated my appetite, I had no choice but to eat what was put in front of me. Thus began my unwilling initiation into the exquisite delights of cheese that it would have been such a pity to have missed out on through sheer childishness. I have Paris to thank for that too.

Dommanget worked at the Royal Belgian Observatory at Uccle, so I went to Brussels to perfect my knowledge of astronomical site testing. I took the opportunity to do some tourism; it was my first time away from Spain and I wanted to make the most of the occasion. Brussels was not yet the European capital and, having just left Paris, I was not in the least impressed by it.

Once back at the French capital, and with the help of Professor Barbier of the Astrophysics Institute, I also took the opportunity to learn about the observational and data reduction techniques for measuring the brightness of the night sky. While working with him in Paris, I made the fateful discovery that we are stardust and accompanied Barbier to Haute-Provence Observatory, where he had his photometers mounted. It was also my first experience of couchettes on trains.

I was awed by the magnificent panoramas from the Observatory onto the Rhône Valley, with the bright, snowy Alps blocking the eastern horizon. It was the first time I had seen a true astronomical observatory (the one in Paris seemed more like a beautiful museum, a bit like the observatory in Madrid, but on a grander scale). Adding a mysterious scientific touch, the white domes blended with the Provençal landscape, which I had previously known only from the Impressionists.

I was bursting with curiosity and keenness. I familiarized myself with each telescope, its instruments, and auxiliary equipment. Everything amazed me and I took copious notes. I overcame my shyness, bad French, and scant knowledge of astrophysics in an effort to speak to whomever would lend me an ear. I kept a tight rein on my prejudices and wanted to learn as much as I could there, which for me was a great deal. I paid endless visits at all hours to learn whatever they could teach me, ferreting out information as far as my timidity and complexes would allow. All this enthralled me and bolstered my research vocation. I reviewed my notes in my room, underlining what had seemed to me to be of particular interest. With my humble scientific background, this observatory seemed to me an achievement beyond our reach.

I had been fortunate in making Daniel Barbier's acquaintance; he was an astrophysicist of repute, a kindly and approachable person, and an authority on stellar atmospheres. Since we spent a lot of time together, we spoke a great deal, and he inundated me with his knowledge of the Universe. It now became clear to me that I should dedicate myself to astrophysics. I now fully understood that the fact that we are stardust was well-proven and much more than a fanciful turn of phrase.

It is humbling to consider that our planet and our conceited species are all dust generated from the stars. The elements that make up our precious bodies were cooked slowly and eventually expelled catastrophically by stars that died,

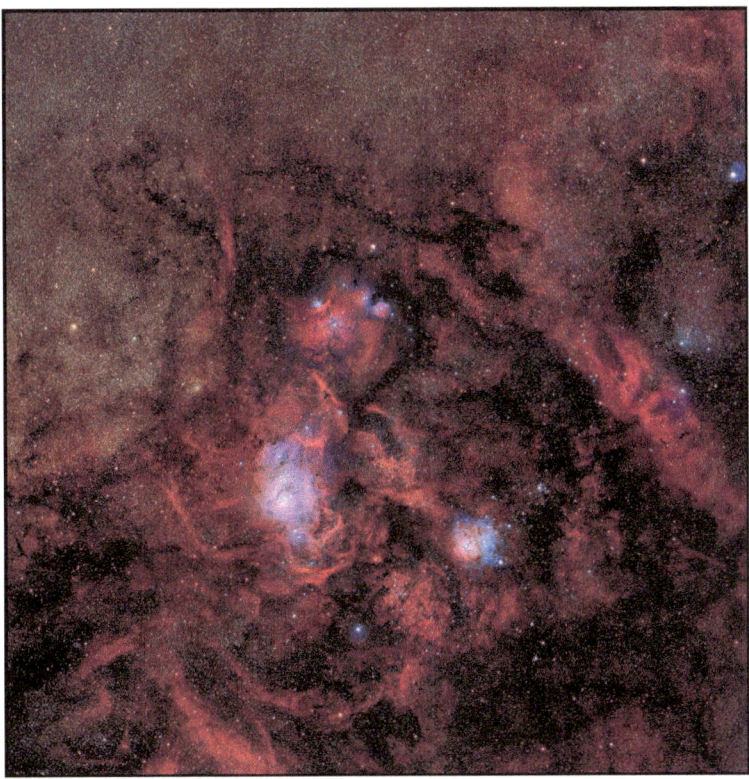

Fig. 4.3 Stars and planets being born in immense clouds of interstellar gas and dust, recycled by dying stars. The image shows the Lagoon Nebula (M8), The Trifid Nebula (M20), and NGC 6559, all at a distance of some 5000 light years. Image obtained with the STC astrograph located at Teide Observatory. (Credit: Daniel López/IAC)

seeding the interstellar medium with the processed matter from which our Solar System was born. We are privileged to share in the knowledge that we are recycled matter, and that this recycled matter which we are will continue the Cosmic cycle to make new stars that will shine bright in a future firmament. We transcend death through evolution, which is nothing more than another manifestation of life, a phase change, to put it eruditely.

The life and death of stars are peppered with violent evolutionary episodes. Stars emerge in clusters after a dramatic birth and meet a catastrophic end that seeds new stars. These violent episodes produce such beautiful and surprising objects as planetary nebulae, novae, supernovae, neutron stars, black holes, and many exotic etceteras more. The stars, then, are factories of all the elements of the periodic table, the true philosopher's stone so eagerly sought after by mediaeval alchemists.

Inside the galaxies, including our own Milky Way, stars are being born in clusters through the gravitational accumulation of interstellar gas and dust. This protoplanetary matter gradually evolves, through endless fusions and collisions, to form suns, planets, and satellites. And on our Earth, it has produced what we call life, which in its turn has continued to evolve until, after many millions of years, it produced the human species. Is that not reason enough to be passionate about astrophysics?

Science has enabled us to grasp a less mythological and more realistic picture of the mysterious Universe to which we belong, together with a new perspective of ourselves. It seems, then, that we are neither more nor less than recycled matter, the ephemeral soot of stellar life. Science has lifted the veil from our eyes. But we still await answers to the most fundamental questions concerning our reality and that of our father, the Cosmos.

5

The Zodiacal Light (1962–1982)

The zodiacal light can be seen from the Earth as a faint pyramid-shaped glow centred on the ecliptic in early dawn or late dusk. It is caused by the scattering of sunlight by tiny interplanetary dust grains in the Solar System. The spectacle of the roseate pyramid of the zodiacal light reaching almost to the zenith, easily seen with the naked eye from the limpid Canarian summits (particularly in the spring and autumn months when the ecliptic is at its most perpendicular to the horizon), is at once beautiful and unsettling. This 'false dawn', as the Arabs called it, is a little-known phenomenon to the public at large because of the light contamination that drowns out this visual boon of nature.

The zodiacal light provided our first clue to the existence of the interplanetary medium and gets its name from being visible in the constellations of the zodiac. The axis of this pyramid coincides closely with the invariable plane of the Solar System ('the maximum plane of Aries', as Laplace dubbed it), which forms an angle of one degree, thirty-six minutes of arc with the ecliptic. Given the morphology of the zodiacal cloud and its nature, it is logical that it should be symmetrical about the sun and be located in the same plane in which the major planets describe their orbits. We may picture it as having a lenticular shape, extending out to at least the orbit of Jupiter, with a central hole occupied by the sun.

Zodiacal dust may be considered simplistically as a remnant of the protoplanetary disc, the residual detritus of collisions among planetesimals of the primordial nebula. How can it be possible that these clouds of primordial electrically charged micro-grains, subjected as they are to significant electromagnetic forces that tend to disperse them, should remain stable over millions of years? This and other questions regarding the interplanetary medium have yet to be answered.

The Very Large Telescope Interferometer (VLTI), observing in the infrared, has recently detected 'exozodiacal light' around 92 stars in the Galaxy, an indication that these stars possess an interplanetary medium, although exoplanets have been found around only 14 of these old stars.

Because of its proximity to the ecliptic, the zodiacal light is more vertical against the celestial vault in the tropics and is best viewed from there. That is why one of the goals of knowledgeable nineteenth century travellers to these regions was to contemplate the zodiacal light. When he came to Tenerife in the mid-nineteenth century, Piazzi Smyth also observed and discussed it in his writings. These days it is not easy to see it, except at very clear sites without light pollution.

I mention all this because it has a great deal to do with the beginnings of astrophysics in the Canary Islands. During my journey of apprenticeship in France and Belgium in 1962, I made the acquaintance of the French astronomer René Dumont, at Haute-Provence Observatory. It was from him that I first heard about the zodiacal light. He told me he was constructing a telescope specifically designed to measure its brilliance, but that it would be impossible to observe it from his observatory in Bordeaux. I immediately set about persuading him to test the telescope in the Canaries. We had no professional telescopes in Izaña, and after my conversion to astrophysics I wanted to commence research right away. We got on well together, and I encouraged him to bring his family on vacation to Tenerife so that he could personally check for himself all the good things I had been saying about the site. He arrived with his family that summer, and I managed to arrange accommodation for them at Izaña Meteorological Observatory, where I was living. Our families became friends and remain so to this very day.

After his holiday, René was won over, to the point that, near the end of 1963, the 'Bordeaux Telescope' had arrived at the docks in the port of Tenerife and was set up at Teide Observatory the following month.

The French expedition, consisting of astronomer René Dumont and engineer Guy Soulié, was funded for only two months. As they were obtaining very good results, I proposed that, since they could not observe in Bordeaux, they leave the telescope in Tenerife, that I take charge of the observations, and that we exploit the data together. They conferred with their director, who accepted the deal, and left us the telescope. This agreement with the University of Bordeaux was to serve as a template for later cooperation agreements that resulted in the Canarian Observatories being filled with telescopes. Our philosophy was, 'We have an excellent sky and you have magnificent instruments, so we shall supply the sky ("buildable land") and you, the telescopes so that we can then work together' (Chap. 10).

In 1964 we began systematic all-sky observations of the zodiacal light with the telescope provisionally set up next to a primitive building that had been abandoned by the meteorologists, where we already had a system for measuring atmospheric transparency. The power supply, measuring systems, and control post were set up in the building. The 30-cm spectrophotopolarimetric telescope was an advanced instrument for its day and was specially designed to measure diffuse sources; today it stands on exhibit in the foyer of the library at IAC headquarters in La Laguna. We had at last managed to acquire a professional instrument for the observatory that we wanted to build in the Canaries. It is worth emphasizing that, with the data from this instrument, we carried out the first astrophysical observations, published the first results, and wrote the first Spanish doctoral theses in astrophysics.

I remember those times in the sixties and seventies when I had unlimited energy for everything: site testing during the day, observing with the Bordeaux Telescope during the long winter nights, studying astrophysics, breaking down barriers, attracting foreign telescopes, introducing the teaching of astrophysics into Spanish universities, creating the Institute of Astrophysics,

Fig. 5.1 Image of the zodiacal light and the Milky Way over Mount Teide. (Credit: (Daniel López/IAC)

etc.—on top of caring for my ever-growing family, which now included my mother and sisters. Now, at my age, it all looks impossible, unreal.

In order to survive, I had to set up a bunk bed in the building under the measuring instruments of the Bordeaux Telescope. I slept there from sunrise till lunchtime so that the children would not wake me. By 1964 we had two children and were still living in the former quarters of the Meteorological Observatory Chief. This odd situation continued night after night. I had in mind what Professor Barbier once told me when we dined at Haute-Provence Observatory: 'There are only two kinds of astronomers, liars and cynics. The former,' he said, 'always make a public show of complaining about cloud preventing them from observing, and the latter pray to be clouded out.'

The excellent results that we were publishing on the zodiacal light also served to prove, with hard data, the exceptional astronomical conditions of the Canarian summits. Our publications also attracted the attention of Professor Ring of Imperial College London, who in the early seventies was installing at our observatory what would for that time become the world's largest infrared telescope. We had agreed that he would send a student of his to study the motion of interplanetary dust particles by means of Doppler shift measurements of the zodiacal light. The student was none other than Brian May, who would later leave astrophysics to become the celebrated guitarist of

Fig. 5.2 The astrophysicist René Dumont and the author next to the newly-installed Bordeaux Telescope in Izaña. (Credit: R. Dumont)

Fig. 5.3 The Bordeaux Telescope on permanent exhibition in the foyer of the Library at IAC headquarters. (Credit: Miguel Briganty/IAC)

Queen. He says that some of the best-known songs of the group occurred to him precisely when he was working at Teide Observatory. Years later, I asked why I had never seen him with a guitar in his hand and he replied, 'Well, I hid it because I didn't want you to think that I was wasting time!'

This young doctoral student had to build everything from scratch, including assembling the prefabricated tin hut in which he was to reside and setting up his measuring instruments. We gave him all the help we could, and I assigned Carlos Sánchez Magro, who was also doing his thesis on the zodiacal light, to take care of him. I remember the Brian May of those days as a gangling, long-haired, but clever, very friendly, and courteous youth. At the time there were a lot of long-haired young men at the Observatory. To his credit, Brian May, in spite of all his success in the music world, has never lost his love for astronomy. He not only brilliantly concluded his thesis in 2007 but is also a driving force for astronomical research from his present prominent position. He engages in outreach with notable distinction.

Around the Bordeaux Telescope there arose what we called at the time the Nocturnal Sky Section, the first organized astrophysics research team in Spain. Sections dedicated to solar physics, infrared astronomy, etc., were built on the same model. The procedure was a straightforward one: insist on collaborating with those constructing the telescope in order to train our own people so that

Fig. 5.4 The 'tin hut', which housed British instruments to measure the Doppler shift of the zodiacal light. In the background, the incipient Teide Observatory as it was in 1971. In the distance can be seen the mast of Televisión Española. (Reproduce from the thesis of Brian May. Credit: Brian May)

we could exploit the telescope jointly. It was under the wing of such collaborations that our own research groups grew.

This nascent research team soon had at its disposal, apart from the Bordeaux Telescope, a semi-automatic double telephotometer (dubbed the 'coffee pot'). It was specially designed to perform the spectral photometry of airglow and was equipped with interference filters centred on the principal passbands emitted by the upper atmosphere. We went on to design new, even more automated instruments. I discuss these in more detail in Chaps. 17 and 18. The IAC's Instrumentation Division grew from these early efforts to develop the skills needed to tackle the design and construction of highly advanced instruments for terrestrial and space astronomy, and later on to face the daunting challenge presented by the Gran Telescopio Canarias (Chap. 21).

But to return to our first investigations. Far from cities, even at sites with low light pollution on moonless nights, the sky is never completely dark. The landscape is always bathed in a tenuous glow that does not produce shadow: it is light from the night sky itself, comprising a mixture of glows of different origins. When a telescope observes from the terrestrial surface it captures a mixture of these different glows: light from the upper regions of the atmosphere (airglow), the zodiacal light (from the interplanetary medium), integrated starlight, light from our Galaxy, and extragalactic light (light from objects beyond the Galaxy). The greatest difficulty in measuring each of these components is separating them from one another.

According to the IAC archives, a total of 260 hours of observations were carried out in 1973 with the Bordeaux Telescope and 720 hours with the double telephotometer. There are thousands of entries of nocturnal sky brightness data. The section was active until 1982, and the researchers in the team formed the embryo of various new IAC research groups. I must confess that I am responsible for taking the decision to disband the team in order to continue to push for the formation of the Institute, a task to which I was now dedicating all my waking hours. It was only in this way that I could avoid a conflict of interests between heading my own research team and directing the entire Institute. I had in effect 'left research'. From that point on it became clear to everybody that their Director now minded only collective interests.

Things were going swimmingly between Dumont and myself. We were among the leading groups working on the interplanetary medium and we completed our theses using data from the Bordeaux Telescope.

We made the fullest use of our advantages; namely, the quality of the Teide sky, an innovative instrument, and a method of observing and data reduction developed by Dumont himself. For that reason, we were able, among other things, to carry out reliable measurements of the zodiacal light over the whole sky, including the gegenschein (a small, extremely feeble luminous patch on the sky diametrically opposite the sun, which moves in synchrony with it). Our empirical model of the brightness and degree of polarization of the zodiacal light are still used today to separate the zodiacal light from other sky glows.

By inverting the integral of the measured brightness in different directions, we were able to determine the scattering functions per unit volume in the zodiacal cloud, and thence values for the density of dust particles, together with the physical and chemical properties of the grains. Scattering functions depend on the size and refractive index of the particles. We thus concluded that the density of the dust cloud decreases in inverse proportion to its distance from the sun, and that the particles which are the main cause of the zodiacal light are approximately micron-sized.

The permanence of the zodiacal light might lead one to think of an immobile and stable cloud of particles: nothing could be further from the truth. To begin with, the measured Doppler shifts make it clear that the dust grains describe orbits around the sun like all other bodies in the solar system. To understand the renewal process, we need to take into account the forces to which these micrometeoroids are subjected to bring about changes in their orbits. These changes include radiation pressure acting in the direction of the Poynting vector (radially outwards from the sun), Poynting–Robertson drag (a resistive force proportional to the tangential velocity of the particles), braking caused by the electrostatic interaction between the solar plasma and the charged zodiacal dust particles, and forces deriving from the existence of the magnetic field in interplanetary space, all of which tend to augment the inclinations of the particle orbits, one component being radial and the other polar. For certain particle sizes some of these processes will force the grains to leave the Solar System in time periods of less than a year. But even the slow effects of gravity, acting over billions of years, are capable of clearing interplanetary space of dust of cosmic origin.

There is no choice but to seek effective processes of particle injection in order to explain the permanence of the zodiacal light (likewise for exozodiacal light). Among the main processes proposed are cometary disintegration, collisional fragmentation of asteroids, capture of interstellar grains, and condensation of interplanetary gas. It is now fifty years since we proposed the existence of and looked for transitory 'trails' or 'scars' in the zodiacal cloud (regions of altered density) to track the effects of both injection and ejection.

The key questions to resolve are the size and nature of the tiny interplanetary particles, along with their origin and evolution, all of which is of great importance for modelling the formation and evolution of exoplanetary systems. This matter is of the greatest interest now that there is a growing suspicion that there are few stars without a retinue of planets, all of which means that zodiacal light studies have achieved a new importance. Of course, to make headway in this topic, improved polarization and spectroscopic measurements are needed at all wavelengths made from the Earth's surface and space. But it is now many years since the author and the IAC abandoned this line of research.

6

Astronomical Site Testing (1961–1974)

As soon as I got to Izaña in January 1961 I immediately got down to work. I had managed to find out, roughly, what astronomical site testing was all about: its purpose was to try to measure atmospheric parameters that would indicate whether or not a site was suitable for astronomy. Specifically, it was necessary to make a statistical study of three parameters: cloud cover, sky transparency, and optical turbulence. The first two are fairly straightforward, and the third deals with the blurring of images when the atmosphere is unsteady. The first thing to do was to make an inventory of the instrumentation available on the mountain and see what could be done with it.

I needed only a few days to discover that Teide Observatory consisted of only two small rooms. One of them was topped with a small dome of wood and zinc and housed two small (13.0-cm and 11.2-cm) refracting telescopes on an old equatorial mount. These had been bought from a Catalan amateur astronomer named Calvet. In spite of my lack of astronomical training, I quickly saw that the mount was badly aligned with the polar axis, the telescopes badly collimated (to track accurately, the optical axis of the telescope must be perpendicular to the declination axis of the mount), the tracking mechanism prone to sudden jerks, the shutter of the camera burnt, a set of eyepieces missing, and so on. To make matters even more unbearable, the dome was so badly fitted that, when it snowed, the telescope object glasses and the dome floor got covered in centimetres of snow, and the power line suffered a drop of 100 volts. The only bibliography available to me was a useless unpublished report by Professor Redman, the San Fernando Observatory astronomical ephemeris, and a copy of the elementary *Popular Astronomy* by the Galician priest Father Aller.

F. Sánchez, *The Rise of Astrophysics in Modern Spain*, Astronomers' Universe,
https://doi.org/10.1007/978-3-030-66426-8_6

Cloud cover statistics could be gleaned from the daily observations performed by the Meteorological Observatory, where we were accommodated. Data were stored there dating from the beginning of the twentieth century. The other parameters were more problematic.

The method of determining the transparency of the atmosphere was the somewhat pedestrian one of estimating by eye the 'limiting magnitude' at zenith. The magnitude limit is determined by the faintest stars that can be seen with the naked eye. This is obviously dependent, among other things, on the visual acuity of the observer. The simple determinations of 'visibility' made in meteorology, based on estimating the greatest distance in kilometres that objects can be distinguished, is too crude a method of gauging atmospheric transparency for astronomical purposes.

Nevertheless, in June 1961, using only these primitive tools at the Izaña Meteorological Observatory, sheer enthusiasm prompted me to write a small report bearing the title 'Preliminary study of sky transparency in Tenerife', with an introduction, four chapters, a bibliography, and four graphs. The study served to provide a preliminary idea of the most transparent months.

Still a fledgling physicist, during my holidays in 1961, I managed to acquire an old Kipp actinometer, property of the Laboratory and Workshop of the General Staff of the Navy, from the Optics Institute, but without an accompanying measurement system. An actinometer is basically just a thermal sensor located at the bottom of a tube mounted on a tripod. I then had the wild idea that the apparatus could be used to determine atmospheric transparency by coupling it to a precise measuring instrument. I set about designing a Wheatstone bridge capable of measuring milliamperes with precision. As it could not be built in the Canaries, I persuaded the Nuclear Energy Council to make it for me. They did it quickly, and by the end of 1961 reliable systematic measurements of atmospheric transparency were being made throughout the day.

The only method we had of measuring atmospheric turbulence was based on observing the diffraction rings around stars and then applying a procedure to deduce the level of turbulence from these observations. In the absence of an atmosphere, or with a totally stable atmosphere, these rings would be perfect, whereas in a turbulent atmosphere they would become blurred. In the latter case, star images become jittering luminous speckles.

To get reliable statistics from the data it was necessary to take measurements in a systematic and standardized way, so I drew up a detailed work schedule, allotting days and hours to specific weekly tasks to be carried out by each member of staff.

Fig. 6.1 The author pointing the actinometer, an instrument for measuring atmospheric transparency. (Caption: F. Sánchez)

In 1962, after returning from my visits to Paris and Brussels with a deeper knowledge of what was needed, I could see that we lacked suitable means for the task of site evaluation. I presented my superiors with a 'Report on the selection of a site for an astronomical observatory on the island of Tenerife'. I detailed the requirements for a site-testing campaign lasting a year and a half: equipment, one full-time and two part-time graduate employees, and auxiliary staff, all within a budget of a mere 277,000 pesetas (1362 euros). It fell on deaf ears.

Zalote, Izeta, and I did what we could. I churned out detailed monthly memoires to the Chancellor and Director, summarizing what had been done, the problems arising, and what was needed. The series of reports ran from January 1961 to April 1966. I continued to send reports at irregular intervals until the University Institute of Astrophysics was created in 1973. In their totality, they form a complete and rather dismal narration of what had been done, what had been requested, and what had been suffered in those early years high up on the ridge of Izaña.

Here are a few extracts of what I wrote in 1963, bearing the title 'Report on problems of personnel, installations, accommodation, and services'. This document provides a very graphic snapshot of the situation in those early days:

> The rooms we are using are the best at the Meteorological Observatory; however, severe drafts enter through the doors and windows. When it rains, not only does the rain stream in through the windows but it also dampens the thick walls of the building, and it can take months for the damp to dissipate. The result is that there are rooms with permanently damp walls throughout the winter. ... The dome housing the equatorial is not suited to the winter climate at this altitude. Snow enters and covers the floor with centimetres of snow. The walls are beginning to crumble and are completely damp. In these conditions the equipment deteriorates rapidly. ... The equatorial needs to be overhauled and realigned by a specialist. Faults and a number of odd things have been detected in the clocks, tracking, lighting, and optics.

Fig. 6.2 Mary Almeida (right) and the author at the doorway the dome of the telescope used for measuring atmospheric turbulence (the telescope had been purchased from an amateur astronomer). The balaclavas were for fending off the freezing conditions. (Credit: F. Sánchez)

Fig. 6.3 The old observatory building in its present dilapidated state stands in stark contrast to the imposing solar towers of Teide Observatory in the distance. (Credit: F. Sánchez)

The report also recommended encouraging visits by foreign astrophysicists, 'Besides the didactic benefit and prestige that would be obtained from such expeditions, perhaps others of a material nature might follow. I feel strongly that this is an aspect that needs careful study, and that the Observatory should be endowed as quickly as possible with a certain minimum of housing units and work spaces to pursue this undertaking.'

I also stressed the urgent need to begin to do some science: 'Nevertheless, it is clear that we can now think about commencing certain observing programmes of interest to modern trends in astronomy and astrophysics.'

In spite of my complaints and pleas the problems remained unsolved. The greatest misery of all was our feeling of abandonment and helplessness: while the Director sat in comfort in Madrid, we were surviving marooned on the summits of the Canary Islands.

My companions Zalote and Izeta eventually left the Observatory. We had not been paid for months and had to depend on our family for support. I even made a pitiful application for a 'graduate loan' in 1963 that makes painful reading. But the truth was that from my second year in the Canaries, when I was scheduled to return to Madrid to defend my thesis at the Optics Institute,

I had already become hooked. I had by then realized that what interested me most was astrophysics; moreover, it was clear to me that the extraordinary astronomical conditions on the island summits needed to be fully exploited. The preliminary results from the atmospheric turbulence and transparency measurements, together with the large meteorological database for Izaña, showed that the site was as good as, or even better than, the sites tested for the European Southern Observatory.

Almost without noticing it or putting it into words, as often happens with deep-rooted concerns, what was to be the essence of my life began to take hold of me: to promote the scientific, technical, and cultural exploitation of the Canarian skies, and establish astrophysics and its related technologies in Spain. These aims have been my lifelong dream, my greatest passion. So we stayed on in the Canaries, and not because I lacked offers and temptations to go elsewhere.

What times those were, when everything had to be reinvented, from astronomical site testing to astrophysical research, without the necessary means or knowledge. In the meantime, my wife and I were building our family on the peaks of Tenerife. As an additional burden, I had to 'serve the motherland' in Madrid as a reserve officer of 'the glorious Spanish infantry' for two long summers. I was also busy with my doctoral thesis, which I succeeded in reading in 1969 at the Complutense University of Madrid. As if all that were not enough, I had also begun to lecture in physics at the University of La Laguna. They were crowded years.

In spite of everything, towards the end of the sixties, we managed to complete the astronomical site testing of Teide Observatory, and the results were published in Spanish in a national astronomical journal. The results caused barely a ripple internationally, but they served to encourage those who were beginning to develop an interest in the Canarian summits. On the strength of our results, we succeeded in attracting others to carry out their own site-testing campaigns (Chap. 10). The results of all our site testing are summarized in my article 'Astronomy in the Canary Islands', published in the journal *Vistas of Astronomy*.

Very important events occurred in our household during that decade: our first son, Jorge, was born in 1962, and a year later Elena, Ana in '65, and Jesús in '68. My father died suddenly in 1967, so I immediately brought my mother and two sisters to the Canaries. I was the eldest sibling and the only man of the family. We had the good fortune to find a flat for them opposite ours in Santa Cruz, where we had just established our family residence. It was wonderful. We looked after one another and no longer felt alone in the Canaries!

Our economic problems and worrisome lack of means continued until the seventies. In 1974 the Observatory became part of the newly formed University Institute of Astrophysics of the University of La Laguna, but was not itemized as such in the university budget. Unbelievably, everything, even the salaries of personnel (when they were lucky enough to be paid), came under a section the University reserved for 'non-inventoried fungible material'. In other words, we were categorized along with paper and other minor stationary items, and—needless to say—we had no national insurance coverage or job security. I remember at the beginning having to take out a private insurance policy; on a high mountain site and with children, safety concerns and future prospects were a cause for apprehension, however optimistic one tried to be.

As the recently created Astrophysics Institute began to become more active, it started to accumulate debts. As its new Director and the first ever Spanish professor of astrophysics, I complained repeatedly to the Chancellor and Manager of the University, and begged them to include the young Institute in the budget. I always received kindly replies, but the matter drifted on in abeyance.

That was the situation when, in November 1974, the Chancellor (Professor Fernández Caldas), having announced that the Minister for Education was coming to visit the University, said that he planned to take him on an excursion to Las Cañadas del Teide. To round off the visit, he thought, it might be an idea for the Minister make a detour to the Observatory. I decided to risk everything by pointing out to the Minister the lamentable state of things. We quickly assembled a report on 'The present and future of astrophysics in the Canary Islands', together with an economic study, accompanied by a budget for ordinary expenditure, including salaries. I even prepared notes for a brief talk. I did not want to leave anything out.

It was a splendid day on the summit, in contrast to the wintry weather at sea level, which was hidden from view by a sea of clouds. Everything got off to a good start. As we went from dome to dome, enjoying the sunny and dramatic spectacle of Mount Teide, I took the opportunity to outline my basic ideas, just in case there would not be time to do so later (as was indeed the case). I informed the Minister with my usual passion that we had all the natural resources to become a major observing centre for the new European astronomy. The cost to Spain would be low but the returns enormous. We had, I explained, already begun discussions with our European colleagues to formalize the setting up of modern telescopes on the Canarian summits, but it was vital to exploit the opportunity intelligently and without delay. Money, I emphasized, was necessary to achieve all this. I also told him that, to prepare for this opportunity, we had already begun to train Spanish astrophysicists.

He listened with great interest. I now held a professorial chair and succeeded in communicating to him my conviction and enthusiasm. Halfway through the visit the Minister, a member of Opus Dei, took me by the arm and said encouragingly, 'Paco, we must somehow make this happen, whatever it takes.'

They invited me to have lunch with them at Las Cañadas Parador. First, we heard mass at the chapel of the Parador, an habitual procedure at the time among Franco's ministers. During dessert the minister said that he wanted to meet with me, the Chancellor, and the Manager of the University, together with his Director General of Planning and Investment, who was accompanying him. I still remember, every time I pass that parador and the place where we met behind closed doors (it was a bedroom). I began by giving a short summary of the many good things that were beginning to happen and I then went on to detail our precarious economic situation. He stopped me short and asked me point blank how much I needed. Thinking of possible

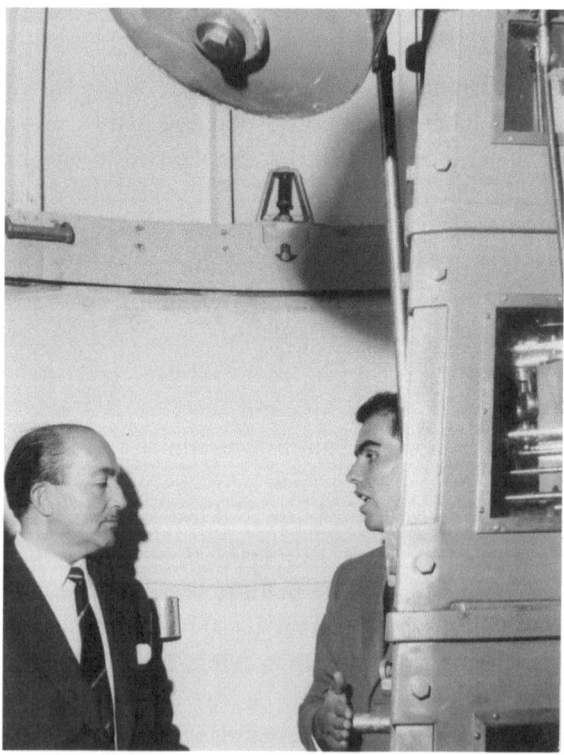

Fig. 6.4 Jesús Rubio García-Mina, Minister for National Education, on a visit to Izaña in 1961 to inspect site prospection activities at Teide Observatory, which he had created by ministerial order. The author explains the use of the telescope for measuring atmospheric turbulence. (Credit: F. Sánchez)

reductions in promised amounts, I hazarded a figure that was double the amount we had calculated: we had prepared a budget of only 5,000,000 pesetas (30,000 euros). He said it would be no problem and ordered the Director General and Manager to include an entry in the budget for the Astrophysics Institute. And that is how, straight of the blue, we entered into the state budget! That is how things were done in those days, when a minister, in November and right off the cuff, could modify the budget for approval by the Cortes in December. At that time the universities were totally controlled by the Ministry of National Education, and ministers were all-powerful.

Of course, all the tribulations I have recounted would finally be amply rewarded by the great successes that were to come. But at the commencement of the astronomical site testing all that lay well into the future. On a personal level, those early years were for me full of discoveries and excitement, both professionally and with regard to my family. All was infused with the adrenalin that pumps when one finds oneself fighting for existence.

7

Astrophysics Versus Astronomy (1961–1972)

In retrospect, the powerful resurgence of research in Spain in the last third of the twentieth century—from zero in some disciplines—is astounding. The decadence of science in our country has its origins in the distant past. It is not my intention here to delve into the causes, which are very well known. What was most lamentable is that we ourselves had begun to believe that we were fundamentally incapable of doing scientific research. We had become content to see ourselves as a nation of artists and writers, a notion borne out by Fernando VII's decision to rededicate the Villanueva building built by Carlos III to further the cause of advancement in the sciences ('Laboratory of Natural Sciences') as an art museum. It is now the Prado Museum. An example already mentioned is Unamuno's famous dictum, 'Let others invent'.

The renaissance of science that occurred in Spain at the beginning of the twentieth century was engulfed in the chasm that was the fratricidal Civil War of 1936 and its aftermath. Some astronomers, for example Pedro Carrasco of the Madrid Observatory, had begun to move towards astrophysics, but exile had extinguished the occasional flame. By the middle of the last century the panorama was bleak. The enforced isolation of the country caused most of the sciences to descend into a state of decay. The great technical advances brought about by the Second World War did not reach Spain. Spanish astronomy had not evolved to incorporate astrophysics and remained where it was at the start of the twentieth century. I remember the sad case of Arturo Duperier and the day of his death at the beginning of 1959, which was declared a day of mourning by my faculty, at which he had held a professorial chair. At the time I was accompanying a fellow student to Navacerrada, where I skied for the first time in my life. During his exile in England, Professor Duperier became an

© The Author(s), under exclusive license to Springer Nature Switzerland AG 2021
F. Sánchez, *The Rise of Astrophysics in Modern Spain*, Astronomers' Universe,
https://doi.org/10.1007/978-3-030-66426-8_7

authority in the growing study of cosmic rays. He was able to return to Spain in the fifties and began lecturing on cosmic rays at the Faculty of Physical Sciences. He persuaded the British to loan him measuring instruments, but—incredibly—he was never able to extricate them from Spanish customs! Such was the state of affairs in our country.

The great effort and tenacity of many people in remedying the decadence into which scientific research had fallen in our country in order to get to where we are at the dawn of the twenty-first century are neither well-known nor valued by society. We may say that we have managed to reach the level of our neighbouring countries in record time, at least in some disciplines, and the story is worth telling. We are concerned here with the story of astrophysics because it is a tale of great significance and because it is the example that I am most familiar with. Frankly, astronomy in Spain had not yet advanced towards astrophysics, but remained fossilized in the state it had been in at the dawn of the twentieth century.

Many of our foreign colleagues speak of the 'Spanish astrophysics miracle' when referring to the unexpected leap forward taken by our country in this branch of science in the final third of the twentieth century. But the reductionist view that that 'miracle' was brought about by the avalanche of foreign instrumentation set up in our country provides no real explanation. There was a much greater avalanche of foreign instrumentation in Chile, but with entirely different results. For me, the key issue was the presence of people who were able to see, understand, and take full advantage of the circumstances. They were many, and it was necessary to unite all their efforts to attain this goal.

At the beginning of the sixties astronomy in our country was poorly provided for and heavily bureaucratized. In practice, it was dedicated to positional astronomy and celestial mechanics; that is to say, the orbits and ephemerides of asteroids, comets, and visual double stars. There were only three chairs in astronomy, all of them occupied by mathematicians. There were three main observatories manned by very few professional astronomers: the National Observatory of Madrid (part of the Geographical and Land Registry Institute), the San Fernando Observatory in Cádiz (belonging to the Spanish Navy), and the geophysically-orientated Ebro Observatory of the Company of Jesus.

The urgent need to improve this lacklustre panorama was clear, and for that to happen many things would need to be changed, but where to begin bringing about such a revolution? It was necessary to proceed with caution and shrewdness since ingrained habits, vested interests, mistrust, and entrenched powers were great obstacles. Convincing strategies, however daring, needed to be devised and effective working attitudes promoted. All this will be dealt with in Chaps. 8, 12, and 15. Here, I emphasize two very specific strategic manoeuvres that served us well.

One of our first strategies was to hoist and proudly wave the banner of astrophysics, presenting it as 'modern astronomy', something new and attractive, and hitherto unknown in Spain. In astrophysics the stars were more than mere points of light having mass, as in the 'old' classical astronomy. In astrophysics they had huge dimensions, were bodies of shining matter with measurable chemical and physical properties, filled with energy, astounding, and in the process of evolving. We set about popularizing and promoting astrophysics with this discourse in mind. Our aim was to raise astrophysics in opposition to astronomy, although it must be admitted that the former is conceptually part of the latter. But the ruse worked, and nowadays, with everything in its proper place we naturally use either term indiscriminately.

As a demonstration of this, in 1975 we convened a meeting ('First National Assembly of Astronomy and Astrophysics) in the Canaries to draw a conscious distinction between astronomers and astrophysicists. In contrast, in 2018 we held a meeting in Salamanca under the title 'XIII Scientific Meeting of the Spanish Astronomical Society' (SEA). If we go to the SEA website, we find: 'The Spanish Astronomical Society brings together more than 800 professional astrophysicists, comprising more than 600 PhDs in Astrophysics and 200 junior members.' It is now no longer necessary to accentuate astrophysics as such.

The posters for the two meetings clearly demonstrate a drastic evolution. The 1975 poster, which, through lack of resources, I designed myself, is simple and minimalist. Contrast this with the professionally produced complex composition for the 2018 poster, with its reproduction of the magnificent Salamancan fresco of the Copernican sky. Also noteworthy is the number of patrons of the Salamanca meeting.

Today it is a mark of prestige to call oneself an astrophysicist, but in my case, as the first person to claim on his identity card the profession of 'astrophysicist', it was an uphill struggle to get it recognized as a profession. Astrophysics has now become accepted; news items often appear on television and in other media concerning the 'mysteries of space and the cosmos'. There are also many astrophysicists in our universities.

A further strategic decision, let me state it bluntly and without embarrassment, was for me to obtain a chair in astrophysics rather than astronomy. At the time you either had a professorship or you counted for nothing, and nobody would listen to you. I was able to verify this immediately in 1972 when, now one of the higher caste, I was called to a meeting at the ministry by the Director General of Universities and Research to 'study matters relating to astronomy and astrophysics'. Also attending the meeting were three professors of astronomy (Torroja, Orús, and Cid), and Padre Romañá, President of the Alfonso X Board of CSIC. I took the opportunity to outline and defend my conception of how we should organize ourselves, both nationally and in the Canaries, in order to derive the fullest benefit from the new means of astronomical observing that continued to arrive in Spain.

My ideas clashed with the way of thinking of the others. Their conception, which was both archaic and centrist, revolved around a main national centre that would control all activity in this field: a 'National Institute of Astronomy'. In that scheme, our Teide Observatory would be simply an observing outpost, all data obtained being exploited in Madrid. It was exactly what the foreign

Fig. 7.1 Poster for the 1975 'I Assembly of Astronomy and Astrophysics'. (Credit: IAC)

visitors had in mind for the Canary Islands, take the data back to the metropolis to do all the science there. I offered explanations and produced arguments in defence of my 'revolutionary' ideas for hours. It was necessary, I insisted, to build new self-governing Spanish organisms, adapted to the new circumstances to meet the new realities, that had scientific and technological capabilities, and with their own administrations. More chairs in astrophysics were also needed. The meeting ended in a stalemate: I had at least succeeded in making them listen to me.

I battled on with my plans. In Spain, as Cela once said, 'He who holds out prevails'. We knew it was not going to be easy; to succeed we had to fight deep-seated prejudices and personal interests. It was difficult to break with the then current model of bureaucratized and centralized national research macrostructures in Madrid. Fortunately, the best sites, those preferred by foreign astronomers for the emplacement of their modern telescopes, were far from capital cities, in no less a place than on the summits of the Canary Islands. In the rest of this book I describe how the 'Spanish astrophysics miracle' was brought about and the sacrifices that were made to achieve it. Chapters 12, 15, and 16 detail what it cost to create the Instituto de Astrofísica de Canarias and endow it with a novel legal and administrative status that would give it agility and self-governance.

Being different and heterodox in Spain has always been problematic, and the singular organism that is the IAC has long been viewed with misgivings, envy, and even animadversion by its enemies.

Another of our main goals has always been to train and equip home-grown astrophysicists and technologists (see Chaps. 6, 8, 15, and 24). I underline this once again because without this permanent infusion of intelligent and motivated new sap Spanish science would wither at the root.

On an optimistic note (bearing in mind the behaviour of those who govern us), I insist that Spanish astronomy (we may now use this term freely) is now in a strong position and plays a leading role. There are excellent Spanish astrophysicists dispersed throughout Spain and the world. They work not only at the IAC but also in universities and other Spanish scientific research centres with no need for their own observatories. The same may be said of the astrophysical technological teams that lend support to the nascent science industry in Spain.

As I never tire of repeating, it is wonderful to witness this and to find articles authored by Spanish astronomers in every issue of the scientific journals of greatest impact in this branch of science.

Fig. 7.2 Poster for the 2018 'XIII Scientific Meeting of the SEA'. (Credit: SEA)

8

Astrophysics Enters the Spanish University Curriculum (1970–1985)

Regulated teaching of astrophysics in Spanish universities was not established until the sixties. A mere three decades later we now find the teaching of, and research into, the subject to be a greatly-valued part of the curriculum at universities all over the country (Chap. 24). University technology teams have also gradually emerged in this branch of science to stimulate local technology. In this chapter I describe how it all started.

When initiating a new discipline the usual practice is to learn from well-established centres where the subject is flourishing. That is how it all began in Spain in the middle of the last century, when young physicists were sent abroad to learn astrophysics. But candidates were initially few and grants scarce.

In the Canaries it was clear to us that there was an urgent need to train Spanish astrophysicists if we were ever to exploit the natural astronomical resource of the island summits. The new telescopes that were arriving needed to be manned by young Spanish researchers, so we began to teach them the necessary skills in situ and send them abroad for further training. We relied on the support of our guest scientists, who began arriving with their telescopes, first-rate European astronomers willing to teach and supervise doctoral theses. We have steadfastly persisted in this approach.

An important step was to convince the Ministry of Education to approve and fund what we called the 'National plan for the training of astrophysicists'. We urged the ministry to establish a University Institute of Astrophysics. We issued a nationwide appeal to recent graduates in physics and carefully selected from among the respondents. Very rigorous courses in the third (doctoral level) cycle of physics were imparted to these first students in October of that same year at the University of La Laguna, with twelve hours of classes per

© The Author(s), under exclusive license to Springer Nature Switzerland AG 2021
F. Sánchez, *The Rise of Astrophysics in Modern Spain*, Astronomers' Universe,
https://doi.org/10.1007/978-3-030-66426-8_8

week and a substantial amount of practical work. The course was radically different from the usual fare at the time and was a success. Many of today's most distinguished Spanish researchers and lecturers began their professional career in that school.

We had previously begun gradually to introduce astrophysics at the university. In 1970 and 1971, I gave two doctoral courses, 'Physics of the interplanetary medium' and 'Solar System physics', at the University of La Laguna. In 1971, I also lectured on the 'Physics of the interplanetary medium' at the Complutense University of Madrid. Our strategy was always to couple astrophysics and physics, in contradistinction to mathematical astronomy (Chap. 7).

The creation of the University Institute of Astrophysics at the University of La Laguna in 1973 was a major advance. The Teide Observatory was eventually to become part of the new Institute, thus uniting all astrophysically related matters to the University. In 1976 for strategic purposes I accepted the title of Vice-chancellor of Research at the University of La Laguna, and the following year the Ministry was persuaded to authorize the bachelor's degree in physics at the University. The corresponding ministerial order was published on 10 March 1978, and on 13 March the programme of studies for astrophysics, as part of the second (master's level) cycle of physics, was approved. All this is easy to summarize in words, but an immense amount of work and effort was involved in bringing it all to fruition. These achievements were important milestones in the introduction of this discipline into the Spanish university system and established important precedents.

A fever to create new universities broke out all over Spain in that epoch, and almost anybody would be hired to supplement the teaching staff. How poorly the first generation of graduates turned out became all too clear. In La Laguna we opted for a top–down approach in creating the Physics Faculty. It meant that we had to fight the prevailing academic prejudice against new ideas, but we finally succeeded. We began to produce PhDs in astrophysics and used these to build up the teaching staff. We then set to work on the second cycle and began to turn out physics graduates specializing in astrophysics. With all this experience behind us we began the first (bachelor level) cycle of physics. The Faculty, thanks to astrophysics, is one of the best in Spain today. Official mentions and certifications abound of the excellence of the bachelor's, master's, and doctoral degrees.

These successes were hard won indeed and required humility in accepting existing reality and the need to seek help. In designing the first programme of studies in astrophysics I wrote to a select group of renowned authorities in astrophysics from all over the world. I sent them a questionnaire including the following questions:

Fig. 8.1 The Central Hall of the University of La Laguna. (Credit: Tamara Tayri Muñiz Pérez, IAC)

- What level of previous knowledge (mathematics, physics, etc.) would you consider to be necessary for undertaking studies in astrophysics?
- If only the final two years of the physics degree could be dedicated to astrophysics, which topics (obligatory and optional) should be included?
- What are the main areas of interest in astronomical research today?

I was gratified and encouraged to find that they all responded quickly in a practical and generous way. I lost my fear of addressing my betters. I have kept their replies as a memento. Their counsels helped us to produce a balanced first programme of studies in astrophysics; they also aided me in preparing the programme I used for my professorship in astrophysics.

The answers were generally direct, straightforward, and clear. They unanimously advocated a solid grounding in mathematics and physics before embarking on astronomy. They placed more emphasis on physics and suggested leaving very advanced mathematics for special topics at doctoral level. Basic classical astronomy, they thought, should be given at the start simply for orientation. I also pored over programmes of study that I had obtained from the most prestigious universities. With all this outside help and much thought, a programme of studies was drafted for the first Spanish university course in astrophysics, which was imparted for the first time in October 1978 at the University of La Laguna, just a year later than we had originally planned. I

Fig. 8.2 The Faculties of the Physics and Mathematics at the University of La Laguna, situated in Avenida Astrofísico Francisco Sánchez. (Credit: Tamara Tayri Muñiz Pérez/IAC)

transcribe below a summary of the programme both to demonstrate its level and because it marked an historic moment for the University. It corresponds to the fourth and fifth years of the physical sciences degree. Year-long courses are indicated by (1) and four-monthly courses by (1/2).

Fourth year

- Astronomy and astrophysics (1)
- Statistical physics (1)
- Electronics (1)
- Instrumental optics (1/2)
- Mathematical methods (1/2)

Fifth year

- Astrophysics II (1)
- Instrumentation and astrophysical techniques (1)

- Particle physics (1)
- Optional (1/2)
- Optional (1/2)

Optional courses

- Theory of communications
- Numerical analysis
- Planets and the interplanetary medium
- Relativity (cosmology)

Students also had to do a research project of not less than six months' duration. Full details of the complete programme, with course details, may be consulted in the Official State Bulletin of 6 May 1978.

I need to stress that, for obvious reasons, we wanted the main focus of the teaching of astrophysics in the Canaries to be observational rather than theoretical. Indeed, the advantages of being able to carry out practicals at the IAC Observatories are currently being enjoyed by other European universities in collaboration with the University of La Laguna.

To give an idea of the astronomy being taught at Spanish universities in that period, here is the programme of studies imparted at the University of Zaragoza by Professor Cid Palacios:

Fig. 8.3 Poster announcing astrophysics courses as part of the fourth year of the physics degree at the University of La Laguna. (Credit: IAC)

- Astronomical refraction
- Aberration
- Transformations through translation
- Precession and nutation
- Proper motion
- Reduction of position
- Time and its measurement
- Optics and astronomical instruments
- Auxiliary instruments
- The theodolite or universal instrument
- Precision instruments (only astrometric) for positional determination
- The determination of position at sea
- The equatorial telescope
- Astronomical photography
- Geographical charts

There were only three fourth-year student for the academic year 1978–1979. The following year eight students enrolled for the fourth year and five more for the fifth year. The numbers are quite high nowadays and the students come from all over the world. The teaching of astrophysics at the University of La Laguna is greatly esteemed and has received a number of commendations. Its attraction, quite naturally, has much to do with the IAC and its famous observatories. It must also be mentioned that the previously mentioned shortage in university teaching staff has long been filled by IAC researchers.

The first chair in astrophysics was established at the University of La Laguna at the beginning of the sixties, a decade before similar chairs were established at universities in Barcelona and Madrid—yet another example of what an uphill struggle it was to introduce astrophysics into the Spanish university system.

No sooner than it was created, this first chair in astrophysics was immediately revoked, owing to the power of a respected professor of astronomy at the Complutense University, who was well ensconced in the regime and happened to be the Director of the Teide Observatory at the time. This sorry tale will be told in Chap. 10. It cost heaven and earth to get this decision overturned so that I could present my *oposición* (examination) for the chair. The *oposición* was one martyrdom more, with behind-the-scenes manoeuvring and pressure exerted on the tribunal. That is how things were done in those distant days.

In 1983 we managed to persuade the state administration that, just as there were medical interns in hospitals, so there should be 'intern astrophysicists' at the IAC Observatories. Since then, a national programme to train young physicists, mathematicians, and engineers has been in operation. At the end of their four-year grants the students must prepare and present their doctoral theses. These interns are integrated into research teams at the IAC under the supervision of a tutor to work at the Institute and its Observatories, with the possibility of stays at foreign institutions, which makes it a highly attractive aspect of the ULL–IAC Postgraduate School.

The intimate connection of the Instituto de Astrofísica de Canarias with its parent institution, the University of La Laguna has resulted in a positive and productive relationship on a firm legal foundation (Chap. 12). In what has been a most creative symbiosis, the IAC's teaching of astrophysics is imparted at the ULL, and the ULL's research is carried out at the IAC. Lecturers and research astrophysicists feel themselves to be part of the *Astrofísico*, regardless of which of the two bodies pays their salaries. PhDs are doubly affiliated to the

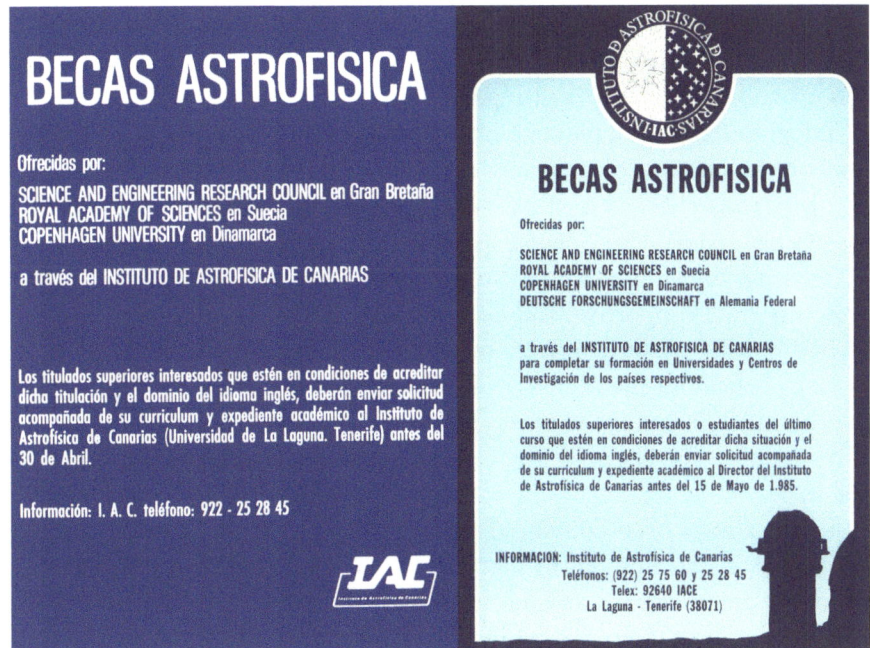

Fig. 8.4 Early PhD studentship posters of the IAC. *Left:* Offer of astrophysics internships. *Right:* Grants resulting from the Multinational Agreement in Astrophysics. (Credit: IAC)

ULL and the IAC. This set-up also serves to counter any tendency towards secession, the natural tendency in our land.

In 1985, after the much-publicized inaugurations of the Institute and its Observatories (Chap. 13), I finally took the decision formally to relinquish my teaching duties at the University, regardless of the consequent substantial drop in my salary; at the same time I requested special leave of absence from my chair in order to dedicate my efforts even more intensely to the cementation of the IAC. It bothered me greatly that, with so much travel and commitments taking more of my time, my students had begun to see less of me in the classroom. As the Spanish saying goes, you cannot be ringing the bells and following the procession at the same time. I was leaving behind many years of university teaching: physics for chemists, engineers, and mathematicians in the sixties, and astrophysics in the seventies and eighties.

Astrophysics research teams gradually began to emerge in Spanish universities, with openings for lecturers at various levels. Graduates and doctors in astrophysics were beginning to appear everywhere. Today there are research teams scattered all over Spain (see Chap. 24).

It is only fair to make special mention of the Department of Theoretical Physics and the Cosmos at the University of Granada (its name has been tempered to the winds of legislative change vis-à-vis the Spanish university system). Thanks largely to its creator, Professor Eduardo Battaner, this department has produced many very well-trained Spanish astrophysicists.

Eduardo Battaner, a pioneer in this discipline, was a key figure in blazing a trail on the summits of the Sierra Nevada during the seventies. That many of our best researchers today have been students of his is proof of this. I have always maintained that Spanish astrophysics would not have become what it is today without this multi-talented teacher, researcher, and science popularizer; this loyal, supportive, and keen-witted friend.

It is not my purpose to raise memorials to renowned Spanish astrophysicists, so I shall not mention here all Battaner's achievements and adventures. Neither shall I recount the stories of the many university research groups that now populate the country. There is, of course, much to tell, but others will tell it.

I have always been convinced that astrophysics cannot flourish in Spain through large research centres alone. As in the most developed countries, also necessary are university research teams. I have tried always to stimulate their creation and help them prosper. As the first doctor and professor of astrophysics in Spain, I served for many years on doctoral and professorial tribunals.

In 1977, when I was Vice-chancellor of Research at the University of La Laguna, I wrote a letter to all the faculty deans with study programmes in physics, explaining the advantages of including astrophysicists among their teaching staff and the opportunities that would ensue to create departments of astrophysics. I addressed university chancellors, directors general, and ministers on this issue.

Today, astrophysicists teach at a growing number of Spanish and foreign universities (Chap. 24). I am sure that this group has been responsible for firmly establishing astrophysics in universities in Spain and beyond. Quantitative proof of this effort is the production of PhDs in astrophysics. According to data provided by the Spanish Astronomical Society, 900 doctoral theses were read in Spain in the period 1969–2013; according to the Postgraduate Teaching Division of the IAC, 347 were completed at the IAC between 1969 and the present. The rate of increase has been gradual but steady, from one (mine) in 1969 to the 43 in 2013. What might the future hold?

9

The Reluctant Chancellor (1976–1980)

The end of the seventies, after the death of the Dictator, was a troubled and decisive epoch for Spain. Europe had recovered from the drama of the Second World War but was now in the throes of the 'Cold War', and the geostrategic situation of the Iberian Peninsula and its archipelagos had assumed a new importance. Both inside and outside the country there were high hopes and worries over what was about to take place here. At the Instituto de Astrofísica de Canarias we were deep in negotiations over the International Agreements concerning Astrophysics, and I remember the great interest of European colleagues when they asked questions about what was happening. Some came to Madrid to see with their own eyes what was going on.

The truth was, we had endured too many years of repression and ideological gagging through the vested interests of the ruling class, and there was a new unindoctrinated youth bursting with ideals, that had begun to stir and effervesce. We had already left our rationing cards behind, the free public health service had begun to function, and 'development plans' had improved the Spanish economy. Families made sacrifices to provide their children with an education, which was then a guarantee of a better life. It was clear that we were emerging from misery and were all eager to help move things forward, even at the cost of having to forget past wounds and grievances. On this social reality was built what was to become known as Spain's 'Democratic Transition'.

On 18 November 1976, the Francoist Parliament effectively committed hara-kiri by passing the Law of Political Reform, which ended what had hitherto been called 'organic democracy'. This law was ratified by popular referendum on 15 December.

F. Sánchez, *The Rise of Astrophysics in Modern Spain*, Astronomers' Universe,
https://doi.org/10.1007/978-3-030-66426-8_9

In that same year, Antonio Bethencourt, who had just been appointed Chancellor of the University of La Laguna, invited me to become Vice-chancellor for Research. I accepted the appointment with enthusiasm, bearing in mind the benefits that this might bring for the recently constituted Instituto de Astrofísica de Canarias and the introduction of astrophysics into the university curriculum. Ministerial authorization was soon given for the creation of the Physics Faculty in La Laguna, which started at the second cycle (fourth and fifth year of studies) with an option for astrophysics. We also managed to get the President, on an electioneering visit to the Canaries, to unblock the signing of the Treaty concerning Cooperation in Matters of Astrophysics, which still continues to give such good results (Chaps. 10 and 11). He also provided funds for the development of research programmes coordinated by the University; with these funds, I managed to ensure the launching of aquiculture in the Canary Islands, uniting the efforts of the University of La Laguna with the Spanish Institute of Oceanography, and the Taliarte Centre of the Local Government (Cabildo) of Gran Canaria.

At that time the University was a political cauldron, with fledgling political parties jockeying for visibility, agitating, and competing in order to stand out from the others. New leaders were learning their trade, and any topic whatever provided a motive for organizing noisy debates and rowdy protests. Francoism was in its final, spasmodic death throes. There was no question of university autonomy: the budget came from the ministry, the minister appointed the chancellor and also named, at the behest of the chancellor, the vice-chancellors, deans, and other academic authorities. There was a daily menu of strikes, assemblies, protest days, stoppages, mural graffiti, posters, and so on. It was also a time of protest concerts, which even reached the University of La Laguna. When Lluis Llach came to sing at our University, Manuel Fraga, who, together with Arias Navarro, ran the Ministry of Governance, banned the concert. As soon as Llach's plane landed, they whisked him from the Canaries on the next flight to Sweden.

It was an interesting, but disquieting, panorama. The Communist Party was legalized in April and the trade unions some days later. The first free elections were held on 15 June 1977, and the Central Democratic Union of Adolfo Suárez won.

On the same day that elections were being held, Channel 1 of *Radiotelevisión Española* aired on the programme *¿Quién es?* ('Who is it?') a report concerning a young astrophysicist called Francisco Sánchez, who was to be found roaming among the peaks of the Canary Islands while doing and saying exotic things. It was a long programme, lasting more than half-an-hour, shown after the mid-day news bulletin and had a wide viewership. They probably chose

ABC

MADRID, DOMINGO 10 DE ABRIL DE 1977

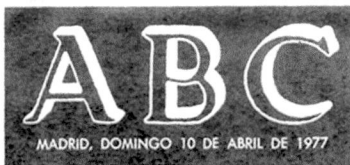

Por resolución del Ministerio de la Gobernación

EL GOBIERNO SUAREZ LEGALIZA EL "PARTIDO COMUNISTA"

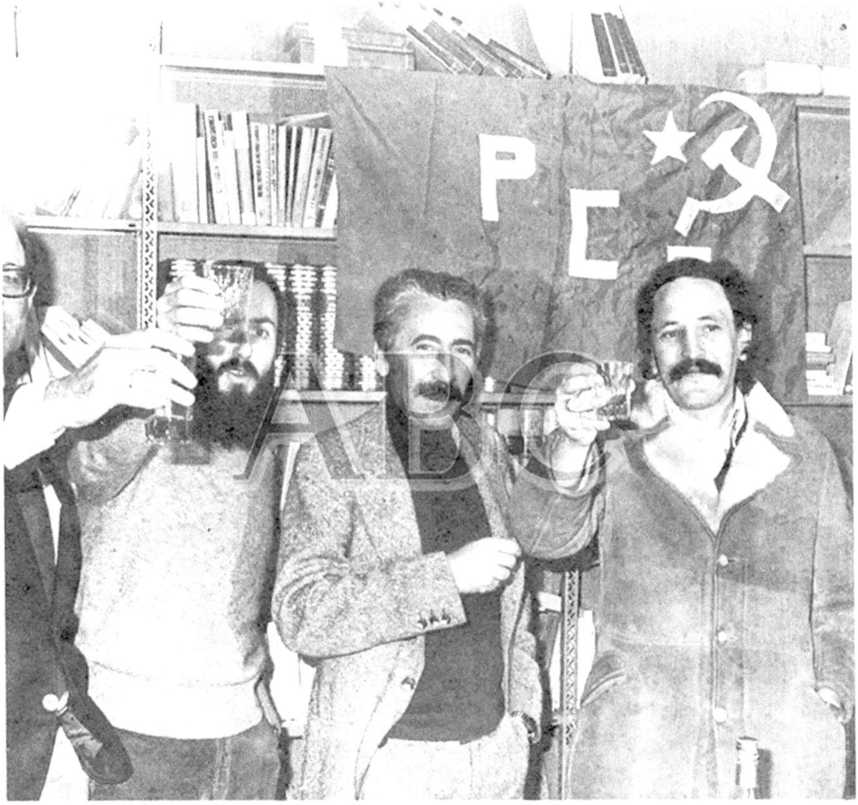

CARRILLO Y LOS SUYOS, EUFORICOS... ¿Y EL RESTO DE ESPAÑA?

Anoche se comunicó oficialmente la legalización del «Partido Comunista de España». Este hecho se produjo por resolucion del Ministerio de la Gobernación. La fotografía recoge a miembros de dicho partido brindando en los locales del «P. C.» de Madrid, minutos después de conocerse la noticia. (Foto: T. Naranjo.)

Ante este hecho, A B C manifiesta en un editorial las razones de su discrepancia

Fig. 9.1 The legalization of the Spanish Communist Party reported on the front page of the national daily newspaper *ABC*. (Credit: *ABC*)

me on that day on the grounds that, in my case, there could be no political connotations. Because of the timing, it was seen by many people who, like us, were anxious for news of the outcome of the elections. It gave great visibility to our astrophysical project.

Two terrible events shook us all in that year of 1977: the killing in Calle de Atocha in Madrid of five labour lawyers in January, and the collision in March of two Jumbo jets at Los Rodeos Airport (they had been redirected from Las Palmas after a terrorist attack alert). To complete the tragic toll, the year closed with yet another doleful occurrence: on 12 December, on the very steps of the University of La Laguna, Javier Quesada, a student, died from gunshot wounds at the hands of the forces of public order.

That was the final straw. The University erupted; indignant and angry students abandoned their lecture halls and took over the University. When the telephone rang at the switchboard callers were greeted with, 'This is the Commune of the University of La Laguna. How may we help you?' The Chancellor took fright and absconded; seeking refuge at the ministry in Madrid. 'He told himself to be gone,' as they say in the Canaries. Faced with this lack of governance, the Governing Council (vice-chancellors and deans) met secretly in the building of the Faculty of Medicine, which was not located

Fig. 9.2 Bouquet of flowers and posters at the entrance of the University of La Laguna marking the fortieth anniversary of the killing of ULL student Javier Fernández Quesada in 1977. (Credit: María Nuria Peña Alonso/ULL)

on campus. With the Chancellor in eclipse and everything in disarray, a decision was taken to appoint an authority figure, an acting chancellor, to take charge of the situation and get everything back in order. They turned to me and I was saddled with the problem. After thinking it over I decided to take up the gauntlet; I felt a sense of obligation and responsibility, quite apart from the fact that I had in any case drawn the short straw.

On 25 January 1978 I took charge of the chaotic situation. I went up to my Vice-chancellor's office on the first floor of the central building, next to the Chancellery, and called the switchboard. I told the Commune that it was the Acting Chancellor calling, and that I was in my usual office waiting for them to come and see me. The office was immediately crammed with people, among them leaders of the embryonic political parties, some of whom would be future presidents of the Canarian Government. They were all talking at once. As there was a great commotion outside the door, I went out to meet them. The great stairway was crowded with noisy students. There was no way to make myself heard, so I took the decision to open the auditorium. We all crowded in, and the enormous theatre immediately became packed. The only lecturer to ascend the platform with me was Gumersindo Trujillo. I could see no other member of the teaching staff there.

It was an impressive spectacle, palpitating and filled to the brim from the stalls to the gods. On the ceiling was depicted a fresco of heroic and victorious youths waving flags and bearing symbols of the Franco regime. The contrast was stark. I managed to establish sufficient calm so that we could make ourselves heard. We debated for about two hours without interruption, everybody having his say. The assembly was becoming calmer, perhaps because it was past lunchtime, until I was able to wind up the meeting and announce that the University would reopen on the following Monday and classes resume.

Peace returned to the University, the Commune disbanded, and lecturers, students, and administrative and service staff resumed their normal activities. I continued to exercise the functions of Acting Chancellor from my Vice-chancellor's office. I had no intention of occupying the Chancellor's office, neither the old one nor the newly built, bunker-like room in the recently constructed central headquarters in the centre of the city. As there was no way that Chancellor Bethencourt would return from Madrid, my position of academic authority was consolidated, and the Governing Council ratified my appointment without the ministry taking any steps to prevent it. It was a strange situation: the Chancellor having fled, the ministry accepted his self-imposed exile as a consummated fact and began to treat me as Acting Chancellor. This situation prevailed until June 1980 when the Chancellor was finally democratically elected. While all this was going on, Professor

Bethencourt in Madrid continued to draw his Chancellor's salary and refused to relinquish his post or put in an appearance at La Laguna. It was not until January 1980 that the minister finally formalized my position as Acting Chancellor, after I had been occupying the post for the two previous years and receiving post from the ministry addressing me as 'The Most Excellent Don Francisco Sánchez, Magnificent Chancellor of the University of La Laguna' (I was quite clearly being castigated for disobedience, as will later become apparent).

They were interesting times. Remember that we were in the middle of the Cold War. Because of my position, I was privy to certain restricted information. I found out, for example, that the Canarian independence movement consisted of two branches. The leftist branch was being funded by 'Moscow gold' channelled through Algeria, and the right-wing element, by 'United States gold' funnelled through Venezuela. The long-standing relation between Venezuela and the Canary Islands is well known. With regard to Algeria, bear in mind that, towards the end of 1975, Morocco invaded Spanish Sahara, and the Saharan people were shamefully abandoned by Spain. And it was to Algeria that both the Saharan refugees and the Canarian supporters of independence fled.

When the political climate changed, the money dried up and the dynamism of the activists diminished until the clamour for independence was barely heard. A level-headed appraisal shows that the Canarians want to be Europeans, not Africans. The islanders have made it clear that they want to be Europeans, and that to be European they must be Spanish.

At the time there existed a short-lived Ministry of Universities and Research, headed by Luis González Seara, a 'democrat of repute' and one of the founders of the magazine *Cambio 16*. Shortly after I had assumed the duties of Acting Chancellor, the Minister summoned me to his office in 1978 to congratulate me for my pacification of the University of La Laguna and for its subsequent smooth running. Next, he announced that he would be appointing me as Chancellor. When I informed him that I did not wish to become Chancellor, and that my only wish was that my successor should be *democratically elected* so that I could return to the full-time task of directing the Instituto de Astrofísica de Canarias, the tone of the meeting abruptly changed. He 'ordered' me to take up the duties of Chancellor without delay; I was handling the situation well, he insisted, and now was not the time to open up the can of worms of democratically electing the chancellor. 'What a precedent that would set!'

But I stood my ground, and he switched from flattering me to berating me; he ended by threatening that, if I refused to do what he demanded, I could forget about the IAC, and that he would put an end to my precious astrophysics institute. I consider myself to be a peaceful and polite man, but the Minister's blackmail made me lose my temper. Some democrat! After telling him that I would follow the dictates of my conscience, I stormed out of his office, slamming the door behind me. The University of La Laguna would eventually become the first Spanish university to elect its chancellor democratically. I, of course, did not present my candidacy; nobody really believed, until the call for candidates expired, that I had no wish to become Chancellor. All this taught me that, in politics, the best way to practise deceit is to tell the truth.

Optimists will say that there is no ill wind that does not bring some good. My slamming of the door at the Ministry of Universities and Research caused the doors of the Presidency of the Government to swing open to the IAC. But that is another battle that I shall describe later.

Building from scratch the process of democratically electing the Chancellor was a long and complicated business. We had opened the Pandora's box of democracy. All the University strata and newborn political parties put forward proposals, some of them quite ludicrous. One of the biggest sticking points was how to weight the votes of the teaching staff, students, and service personnel. Much dialogue and cooling of tempers were needed. One had to brook insults, defamation, and even unrest. All were eager to practise democracy and impose party policies. The walls would often be covered with satirical posters showing the Acting Chancellor dismounting a donkey. But all the obstacles were finally cleared and Professor Gumersindo Trujillo became the duly democratically elected Chancellor of the University of La Laguna in the first ever election of a university chancellor in Spain.

It was a source of great satisfaction to me to see the Law of University Reform of 1983 (which had provoked so much controversy and political unrest in its passage through Parliament) sanction the democratic election of chancellors in all Spanish universities in the same way that this had been achieved at the University of La Laguna three years earlier.

10

A Multinational Astrophysics Treaty (1960–2014)

After we realized that the Canarian summits hosted exceptional conditions for astronomical observing, we immediately set about broadcasting our results in order to persuade those with new telescope projects to come and check these excellent conditions for themselves. Their telescopes and knowledge were vital to the fulfilment of our dreams. In the Spain of that epoch, thinking about getting money to build a telescope of our own that would be anything like competitive was simply a pipe dream, which made it all the more necessary to attract foreign instruments and knowhow if we were ever to initiate astrophysics in our country. Here, I relate how we succeeded in formalizing through an international treaty the arrival of international astronomical instrumentation.

Although our observations and data gathered during the seventies clearly demonstrated the high astronomical quality of the Izaña area, it was essential that these findings be independently verified by other site-testing campaigns carried out by institutions from other countries. It was also necessary to make comparisons of all the islands of the archipelago with summits above 2000 metres and to cover the entire visible, near-infrared, and ultraviolet regions of the electromagnetic spectrum.

Right from the start of my international campaign I found that our foreign colleagues were reticent on two points: dust from the Sahara Desert and the Franco regime. It was a difficult task to overcome their hesitancy on both these issues. For our part, all we could do was to demolish, with data, the myth of harmful Saharan dust, so we added to our measurements of atmospheric transparency and optical turbulence data on dust in suspension.

© The Author(s), under exclusive license to Springer Nature Switzerland AG 2021
F. Sánchez, *The Rise of Astrophysics in Modern Spain*, Astronomers' Universe,
https://doi.org/10.1007/978-3-030-66426-8_10

Without further ado, I extracted the usable data from the abundant archives of the Izaña Meteorological Observatory, which date from the beginning of the twentieth century. I concentrated on the previous twenty years (from 1947 to 1966) in order to overlap them with the measurements we had been taking. The results were quite clear: on only 2.2 per cent of clear days was there any appreciable presence of dust in the atmosphere above 2400 metres. We reported in a publication of 1970: 'From November to February, both inclusive, the number of days without cloud but with dust is less than 1%, and even in the worst summer months the value does not exceed 10.5%.'

In spite of the weight of these data and the fact that we have been accumulating more and more observations for over fifty years, all concordant with our 1970 paper, there are still those who argue against setting up more telescopes in the Canaries, their case still being based the myth of blankets of Saharan dust descending on the IAC Observatories. Again and again, it has been necessary to counter energetically these lies and biases with data. It must not be forgotten that there are powerful interests at play where large scientific structures are concerned. Our latest examples of myth-busting efforts include the European Extremely Large Telescope (E-ELT) and the Cherenkov Telescope Array (CTA): we lost the battle for the E-ELT and were victorious with the CTA (see Chap. 22).

The first to decide not to bring their telescopes to the Canaries were colleagues from the former Federal Republic of Germany. An expedition from Heidelberg Observatory came to Tenerife in 1960 to examine the astronomical suitability of its mountain sites. Their brief stay coincided with a Saharan dust invasion, and they immediately concluded that the Canaries were unsuitable because of dust in suspension. They did not trouble themselves to see what might happen during the rest of the year. The Heidelberg Observatory team settled for the Sierra de Filabres in Almería, where the net astronomical observing conditions are poorer and ironically, as in the rest of southern Europe, there are occasional dust invasions.

It is timely to note here that the first international agreement permitting the setting up of a telescope in the Canaries was signed in 1965 with the University of Bordeaux. Thanks to this agreement, Teide Observatory was blessed with a photopolarimetric telescope (now known as the Bordeaux Telescope, on exhibition today in the vestibule of the IAC Library in La Laguna). The first Spanish astrophysical research team was nurtured with this telescope. This agreement was to serve as a template for all subsequent accords: a foreign scientific institution sets up its telescope at one of the IAC Observatories, Spain provides extraordinary observing conditions and basic infrastructure, the allocation of observing time, etc., and the agreement is signed.

Astrophysicists from Imperial College, London, after measuring the good behaviour of the atmosphere for observing in the near-infrared, requested permission to install their 1.52-metre telescope (optimized for this wavelength range) at Teide Observatory. The corresponding agreement, following the template of the Bordeaux agreement, was signed, and a young Spanish astrophysicist, Carlos Sánchez, was appointed to work with our British colleagues. That is how infrared astronomy, which was later to chalk up success after success, began in Spain (Chaps. 17 and 18).

Another agreement was negotiated in the early seventies with Belgium's University of Mons to install a 60-cm photometric telescope. Since there was no money available, the building to house the telescope was raised by our own astrophysicists. There is now a spectrograph mounted on the telescope, and university students use it for their practicals.

At around the same time, an agreement was signed with the Kiepenheuer Institute at Freiburg set up a 70-cm vacuum solar telescope in a solar tower built at the expense of the cash-strapped University Institute of Astrophysics. We sent another of our young researchers, Manuel Vázquez, to Germany for training in solar physics.

In the mid-seventies an agreement was reached with the University of Birmingham to permit the setting up at Teide Observatory of their pioneering solar instruments based on the principle of resonant scattering. We began by assigning to them the now customary young researcher. His name was Teodoro Roca. That is how the IAC's helioseismology team was born. Today, Teide Observatory is a node of many worldwide networks of this branch of solar physics.

I must now mention a failure. At the beginning of the seventies, French astronomers were building their giant 3.6-m telescope, and they knew that their country did not provide the requisite conditions for its emplacement. They knew of recent site-testing campaigns, including those of the European Southern Observatory, but the very good results we had been getting at Teide Observatory, particularly from our observations of the zodiacal light, seems to have encouraged them to think of Tenerife. They set up an equatorial telescope to measure the optical quality of the atmosphere, and a team of engineers assessed the costs of building their telescope here. Until, that is, the arrival of Professors Cayrel and Jefferies.

From the letters we exchanged, I could see that they were very decided to come to Tenerife. Only two things worried them: the double dependence of the Teide Observatory on the Universities of La Laguna and Madrid, and having to negotiate with the Franco government. I do not know exactly what happened, but they finally opted to take their telescope to Hawaii. General

De Gaulle's celebrated visit to Canada at the end of the seventies had cemented relations between France and Canada, with the result that the Canadians entered into participation in the large French telescope, eventually named the Canada–France–Hawaii Telescope, which saw first light at Mauna Kea Observatory in 1979.

At that time the British were looking for a suitable site for their Northern Hemisphere Observatory (NHO). The proposal for the creation of the NHO dated from January 1967, the year in which Queen Elizabeth II inaugurated the nation's large (2.5-m) Isaac Newton Telescope at Herstmonceux (Sussex), where they had not long before transferred the Royal Observatory, Greenwich, later to be renamed the Royal Greenwich Observatory (RGO). Fred Hoyle, a keen proponent of the NHO, wrote in his autobiography, 'When I asked Walter Baade what he thought of Herstmonceux as a site, he grinned in his pixieish way and said, "It is like building a submarine in the Arizona desert."' There is even a reference to this observatory in one of his famous science fiction novels (*The Inferno*), a copy of which he signed for me many years later during his visit to the IAC. Our British colleagues would later see fit to close the telescope at its Herstmonceux site in 1979 to send it overseas to La Palma, an enviable demonstration of the proverbial British pragmatism: they had the capacity to change their minds unashamedly and relocate their national telescope to a site that was clearly better a mere few years after the telescope's inauguration. They never cease to astonish me.

In the summer of 1971, the British approached the Director of Teide Observatory (the Chancellor of the University of La Laguna) for permission to begin astronomical site-testing in the Canaries. At the end of 1972, J. Alexander of the RGO, after having studied possible sites in the Mediterranean area, wrote: 'The ideal solution would be an international observatory on La Palma'. In 1973 they put out feelers to the Spanish government regarding the setting up of a British observatory on La Palma but with negative results (relations between Spain and the United Kingdom were going through a rather rough patch at the time). They considered Madeira and Hawaii but, since they were truly set on the Canaries, they decided to move towards the idea of an international, or multinational, observatory on La Palma, for which purpose they would need to rely on the recently established IAC (Chap. 15).

In the early seventies, solar physicists associated with the Joint Organization for Solar Observations (JOSO) also decided to come to the Canary Islands to carry out astronomical site testing. They were familiar with the atmospheric conditions in the Izaña area, where Teide Observatory already hosted a solar telescope, so they decided to make comparative measurements on the island of La Palma. They transported their instruments to the summit by mule and set up tents for the campaign.

I wanted to take a look for myself at the summits of that island, where there was not yet any light pollution; I had always thought, given their altitude and location, that they should possess astronomical conditions very similar to those measured at Teide Observatory, so in the summer of 1973 I organized a trip to the summits of the Caldera de Taburiente and encouraged my elder son, Jorge, then a boy of eleven, to accompany me. A car provided by the La Palma Cabildo ferried us to the end of a forest road and deposited us near Pico de Las Nieves. We climbed to the summit as best we could, leaving behind shreds of clothing and skin among the thickets of low *codeso* (laburnum). We pitched a mini-tent that I had bought for Jorge in France and prepared supper on the edge of the caldera, the dark chasm yawning at our feet. We were completely isolated, floating between land and sky, with the firmament pressing down on us. We were a for long time enraptured and overwhelmed before the beauty and unsettling mystery that enveloped us. That sky filled with still, unblinking stars was evident proof of the astronomical quality of the peaks of La Palma. It was wonderful. Was it then that Jorge discovered his strong astronomical vocation, or had the seed already long ago been planted in his subconscious when he was a very little boy and tried to play with the astronomical site-testing instruments in Izaña?

In the morning the caldera was bright in its splendour, with its unfathomable gorges, separated by sharp crests covered in pines that looked just like bushes in the distance. A sea of clouds was beginning to cover the depths of the caldera, looking like steam from a seething, gigantic cauldron. We had a bite to eat and left for Roque de los Muchachos, which loomed in the distance. We continued along the rim of the Caldera de Taburiente, slowly crossing successive peaks, climbing and descending along the edges of precipices with sheer drops of up to a thousand metres. It was exhilarating. We stopped to eat on the way near the Pared de Roberto (Roberto's Wall), a huge basaltic dyke, a Cyclopean wall with a narrow pass, through which ran a path. We reached the JOSO solar physicists' camp close by Roque de los Muchachos in the afternoon. We enjoyed a friendly supper together. They were happy with the data they had been accumulating and told us in detail what they had been doing. Next day, we went down to Puntagorda with the mule drivers, who had brought up victuals.

As a consequence of the good results obtained by the JOSO expedition, the Fraunhofer Institute at Freiburg and the University of Göttingen made a formal application to carry out site-testing campaigns in Tenerife and La Palma, with a view to transferring their present solar observatory on the island of Capri to the Canary Islands and installing telescopes of the latest design.

At about that time Professor Kiepenheuer, Director of the Fraunhofer Institute, took it into his head to build a solar tower on the very peak of Mount Teide. He was convinced that up there the conditions for observing the sun would be excellent, quite forgetting about the turbulence caused by heat from the volcano and high-mountain katabatic winds. A project was planned and reached the stage where material was taken up to the peak by mule and work started. There were so many difficulties with the building work, and such was the growing opposition of the people of Tenerife, who were most unhappy at seeing such contamination of the very peak of their totemic mountain, that the project—thank heavens!—was eventually abandoned. All that remains of this fantasy are a few buried piles slowly being eaten away by sulphurous vapours.

After carrying out site testing on the island of Capri and Sicily, southern Greece, and southern Portugal, the Germans finally decided that the Canarian summits were their best option. They requested permission to build a tower to house a solar telescope on an esplanade close to the two rocks that give Roque de los Muchachos its name. That space is now occupied by a small parking space and a geodetic marker. The corresponding agreement was signed in 1977 and the telescope was operational the following year. After comparative measurements were made, it was finally decided that Tenerife would be the marginally better location, so they had to demolish the tower before transferring their solar telescopes to Teide Observatory.

The interest of our European colleagues in the Canarian summits began to grow. Once the installation of major telescopes in the Canaries had really begun to be taken seriously, the question arose of juridical guarantees through binding agreements. We all agreed on the necessity of a sound legal grounding to guarantee and safeguard the emplacement and continuity of the telescopes. Negotiations intensified and the respective governments were informed.

The British invited me to see British astronomy for myself and they prepared a truly impressive welcome for me. They took me to the Royal Observatory, Edinburgh, the new Royal Greenwich Observatory in Sussex, the Cavendish Laboratory, where, not long before pulsars ('*puls*ating st*ars*') had been discovered, and I met the Nobel laureate Antony Hewish. We dined at Trinity College—in the same dining hall frequented by Newton, they told me. The tour ended in London with a dinner at the Royal Society, with all the splendour and decadent ritual of that institution. We were even served by an old waiter, as in the films. If it had been their intention to dazzle and overwhelm me, then they certainly succeeded. It was clear that we were having humbly to negotiate with an astronomical superpower.

To quell the inferiority complex that was beginning to take possession of me and continue to negotiate with dignity and force, I fell back on a pedestrian, but apt, simile. Circumstances had made us the *nouveaux riches* of astrophysics, rather as the discovery of oil deposits had made the Arabs dollar-rich. The *nouveaux riches* lack the refinement of 'old money' and do not possess the effortless grace of earlier wealthy generations. In my case, unlike my British hosts, I did not speak Oxford English, I was not a famous Cambridge professor, a member of numerous scientific societies, or a Nobel laureate.

I mention one incident to illustrate my situation. In January 1966, I gave a presentation at a symposium (*Zodiacal Light and Interplanetary Matter*) of the International Astronomical Union in Honolulu. At the end of my talk the session chairman came up to me and congratulated me on my temerity. I had dared to speak in English, having no idea of the language. Fortunately, I was fully aware of my limitations and had previously had a summary of my talk translated, which I had distributed prior to taking the podium. That contribution was cited for a number of years.

There was no choice, then, but to face up to the situation. I resolved that Spanish astrophysicists of the future, would no longer feel themselves to be *nouveaux riches*; they would instead be knowledgeable, know how to behave with scientific decorum, speak correct English, and know how to mingle, free from complexes, with their colleagues worldwide. This notion comforted me, and gave me the confidence and strength to negotiate on an equal footing. They also stiffened my resolve to prioritize the training of well-prepared young astrophysicists.

It was time to start profiting from the skies of the Canarian summits as a publicly-owned 'natural resource'. As a rare and desirable resource, it could be neither given away nor auctioned off. It was not ours, but the property of Spain, and we could not allow ourselves to fall prey to frivolity or selfishness; it had to be made available for scientific exploitation to the benefit of science, which is universal, while also remaining a national treasure.

In December 1974 the Chancellor of the University of La Laguna convened a meeting of celebrated astrophysicists with whom we had been in contact from the United Kingdom, Germany, Switzerland, and Denmark. Also invited were the Presidents of the Cabildos of Tenerife and La Palma in order to ensure proper local representation. The meeting was supplemented with flights in light aircraft over the Caldera del Taburiente and Roque de los Muchachos. Everyone was impressed with the savage beauty of the caldera, and the space surrounding the Roque that was available for telescopes was in full view. This meeting was an important milestone.

Also in 1974, Britain's Science Research Council approved the endowment of the Northern Hemisphere Observatory with three telescopes, one of these being the Isaac Newton Telescope. It also gave the Royal Observatory, Edinburgh the task of completing the astronomical site testing of the summits of La Palma. Their exhaustive and thorough examination provided abundant comparative data for various sites and provided compelling evidence in helping them decide that Roque de los Muchachos was where they should set up their telescopes.

The results of this, and all the other, site-testing campaigns carried out in the Canary Islands are summarized in an article of mine published in *Vistas in Astronomy* in 1985.

The added advantage of the lack of light pollution in La Palma resulted in a consensus that Roque de los Muchachos was the preferred site for the emplacement of nocturnal telescopes, given that Teide Observatory was contaminated, mainly from illumination in the Orotava Valley. Negotiations therefore centred on a dual scenario: Teide Observatory for telescopes less susceptible to light pollution, and the future Roque de los Muchachos Observatory for all others. Serious logistical concerns remained: Teide Observatory could be accessed by road, and had electrical power, telecommunications, etc., but a great deal of money would need to be spent to get the Roque de los Muchachos area ready, including roads and other infrastructure.

On 20 November 1975, while we were in the middle of these negotiations, General Franco died. Some of our colleagues went Madrid to check for themselves whether Spain would descend into chaos. On the 22nd of that same month Prince Juan Carlos was crowned King Juan Carlos I of Spain. In December 1976, the Law of Political Reform, which brought the Franco regime to a close, was passed by referendum. The first democratic elections were held in June 1977 and the Union of the Democratic Centre, led by Adolfo Suárez, won the election. In 1978 the new Spanish Constitution was approved by referendum. Spain was now a democracy, and the skies above the Canarian summits were excellent, as the Europeans had verified for themselves. In the face of these realities, the Europeans stopped worrying. All was going well.

There were many meetings between 1975 and 1978, not only in the Canaries but also in Madrid and London. These meetings involved the President of CSIC, Primo Yúfera, who was present at all of them. The first drafts of the future agreements were compiled and sent to the ministries of the exterior of all the countries involved.

Our aim was to articulate good, fair, and lasting agreements that would smooth the path for the arrival of modern telescopes from all over the world to our observatories. We had seen how, in other countries, the establishment of foreign observatories had failed to generated autochthonous modern astronomy so that only a select few of the local astronomers managed to achieve an international reputation. We were not prepared to let the same thing happen in the Canaries. Given that we were fortunate enough to have on our territory such attractive sites for observing the Universe, we felt obligated to obtain the maximum benefit from this publicly-owned natural resource. The concepts 'public' and 'national' were the deciding factor in our negotiation strategy. We were adamant that the telescopes should be set up in wholly Spanish observatories in terms of land, infrastructures, laws, etc.

The initial approach of our foreign colleagues was the one they had adopted in South Africa, Australia, Chile, and even the south of Spain: purchase land and obtain permits to build their own observatories—in other words, genuinely self-sufficient foreign enclaves, using local manpower strictly for auxiliary services. Observatories were to be mere observing outposts in foreign parts, where everything—even toilet paper—had to be imported (that is no exaggeration, the British in La Palma initially imported it from Britain). Observations would be planned in the country of the owner institution, and the data taken back to the metropolis, there to be analysed and published. A purely colonialist concept: extract the natural resource at the lowest possible cost, take it back home for manufacture, and retain the benefits.

We began the negotiations with our anti-colonialist principles firmly in mind. We aimed high: we aspired to a treaty at international level among national governments that would provide safeguards and stability for cooperation over long periods of time. We were also clear about what we were not prepared to accept. We knew that time was on our side, for at that precise moment our partners needed to formalize agreements as quickly as possible in order not to lose the funding they had acquired for finishing their telescopes, buildings, and domes. But neither could we be over-insistent for fear that our partners might have second thoughts and decide to go elsewhere.

In 1979 the International Astronomical Union celebrated its triennial general assembly in Montreal, and the British took the opportunity to convince the Dutch and Danes to join their project, and also to coordinate better with the Swedes. From then on, it was the British who dominated the foreign side of the negotiations.

While at Montreal, I dedicated my all efforts to persuading Admiral Orte, Director of the Royal Naval Observatory in San Fernando, that his observatory should, on behalf of Spain, take charge of the meridian circle that the Danes wanted to set up. I had a hard time convincing him because of his fear of not being up to the job and letting Spain down. In fact, they have done an excellent job and ended up overseeing this transit telescope. They later overhauled a similar telescope they had in San Fernando and set it up in Argentina so that they could cover both hemispheres, thereby converting the Spanish observatory as the world centre for ephemerides and catalogues until the advent many years later of satellites tasked with this activity.

Another of our strategies was to recognize the need to ensure that the observatories we dreamed of building should be well received locally, that the peoples of the islands should understand that the white domes that would begin to mushroom on their mountaintops would bring benefits for all. That is why we spoke (preached, I should say) endlessly of the many benefits that intelligent exploitation of this exotic natural resource could bring to the locality. And slowly, very slowly, the Canarian population—even the politicians—began to understand our argument. It gradually dawned on them that, even if a clear and pure sky is not quite petroleum, it is still a rare resource of great value that could bring genuine benefits locally.

Returning to the negotiations, in order to ensure fulfilment of the agreements and the Spanish character of the observatories, we wanted to take charge of the essential services: power, telecommunications, water, accommodation, restaurants, and access roads. Our politicians finally took this on board, and it is stated in the agreement that Spain's contribution was to assume the costs of these infrastructures, which would be directly supervised by the IAC. That is why both Observatories are run, not by a director, but by an administrator.

But it was politic to give the telescope owners some kind of say in the administration of the Observatories. Apart from anything else, they had greater experience than we did in the management of large observatories, and this would be acknowledged in the treaty through the creation of an International Scientific Committee (CCI, after its Spanish initials).

We also had to persuade our foreign colleagues that it would be helpful for them to hire Spanish astrophysicists with a sufficiently solid background in science and technology to work alongside of them at the telescopes, and not just as auxiliary personnel. That would save the time and cost involved in operating their telescopes on the colonial principle, with their complete, self-sufficient teams from the home country. That approach, along with hiring local staff in strictly auxiliary roles, would lead to a conflictive disparity in salaries, the visitors being exceptionally well reimbursed for working overseas and the islanders receiving local rates of pay. In the end our overseas colleagues had to toe the line and began to employ Spanish staff at all levels. Indeed, for some time now the Director of the Isaac Newton Group has been Spanish.

Another huge challenge was building the difficult access road to the summit of Roque de los Muchachos. In 1976 the National Institute for the Conservation of Nature (ICONA) began work on the road with some of its own funds and partial subsidies from the Association of Cabildos and the La Palma Cabildo. The route was laid out by the mountain engineer José María Galeán of ICONA. I accompanied him in order to ensure, among other things, that car headlamps would not in the future dazzle the telescopes. I have fond memories of Galeán's camaraderie, professionalism, and sheer grit as he climbed and descended the hills and passes, dragging his crippled leg.

Then there was the controversy over the British firm Freeman Fox, which was finishing the construction of the Bosphorus Bridge in Istambul. The British had consulted this firm regarding the best route for accessing the Roque. Freeman Fox issued the Olympian pronouncement that it would be impossible to access the summit from the eastern side; instead, it would be

Fig. 10.1 Scientists involved in the negotiations concerning the International Treaty on Astrophysics ascending towards Roque de los Muchachos on La Palma. Professor De Jager is seen clinging to a horse's tail to help him during the steep climb. (Credit: IAC)

necessary to skirt Los Llanos de Aridane and enter from Garafía in the north. Unconvinced, Galeán and his crew ascended from the north-eastern side of the island and crossed the dramatic Degollada de Franceses, just where the Barranco de los Hombres (a steep gorge descending right down to the sea) meets the Barranco del Diablo (another gorge overlooking the Caldera de Taburiente on the other side). Using a mechanical digger and an expert operator from ICONA, they built the road along a knife-edge with sheer precipices of unstable rock on either side. There are spine-chilling photographs of the work: it was an epic feat. The road works were completed in 1978, and the breathtaking views from Los Andenes are today a great tourist attraction. A viewpoint has been built there for tourists to stop, admire, and take photographs of the view: on one side the Caldera de Taburiente, and on the other a gorge with a vertiginous drop to the sea.

As soon as the road was finished, military personnel used it to find a site for a powerful radar system. They arrived at Roque de los Muchachos and decided that this would be the perfect spot for it! As I shall describe in the next chapter, they very nearly prevented the building of the Observatory and the signing of the treaty. Also in that chapter, I describe how I locked horns with President Suárez over this matter.

As the negotiations progressed, we pushed forward with the infrastructures. In March 1977 we held a meeting with our colleagues in the huts of the IAC in La Laguna to discuss joint planning for the future Roque de los Muchachos Observatory, and in December the Governing Board of the IAC approved a budget of 12,000,000 pesetas (72,000 euros) for the preparation of projects for the Observatory construction works. We wanted the projects drafted beforehand in order to negotiate the building of the Observatory with the government. In 1978, we persuaded the Council of Ministers to earmark 467,000,000 pesetas (2,802,000 euros) for the building of the Observatory, to be included in the Programme of Investments in the Canary Islands, following a visit to the Canaries by President Suárez.

All through 1978, adjustments were made and points of conflict resolved concerning the treaty. We finally got what we wanted: ample, free-of-charge observing time and genuine help in training personnel and building an advanced research centre. Our much longed-for IAC was to be a competitive research centre, where astrophysicists and technologists would be trained, scientific instrumentations developed, and support given to Spanish astrophysics. It was also to be an outreach hub for astrophysics.

The finalized treaty was structured in three levels: 1) an inter-state AGREEMENT, 2) a PROTOCOL among 'designated institutions', and 3) specific ACCORDS between the IAC and each of the 'user institutions'. In the summer of 1978, levels 1 and 2 were agreed and signed. Spain was to assume the initial costs for the access road, Observatory building permissions, power lines, water supply, telephones and telex, the building of a residence and restaurant, maintenance services, workshops, laboratories, administrative offices, and other services.

Level 3 was more specific and required further detailed negotiation. It was not finalized until January 1979. I remember that we met in the huts where the IAC was stationed at the time. It took us three days to revise the texts and we finally signed them on 21 January. We celebrated with Spanish *cava* and we all signed the bottles. I still keep one of them as a prized and significant souvenir.

The final sprint was very Spanish in style: on 10 May 1979 the Council of Ministers authorized the three levels of the multinational treaty at the request of the Ministries of the Presidency of the Government, Universities and Research, and Defence (this was the last ministry to participate (see Chap. 12). On the 26th of that same month the documents were signed during a solemn ceremony at the La Palma Cabildo by ambassadors from the respective countries and the Minister (Pérez Llorca) of the Presidency of the Government of Spain. The Official State Bulletin of 6 July published the documents, which were later ratified by Parliament.

Fig. 10.2 The scientific negotiation team celebrating with *cava* the finalization of the negotiation for level three of the International Treaty on Astrophysics at the old headquarters of the IAC. (Credit: IAC)

Fig. 10.3 Signing of the International Treaty on Astrophysics at the Santa Cruz de la Palma Cabildo by the Minister of the Presidency of the Government (Mr Pérez Llorca) and the ambassadors of the signatory countries. (Credit: IAC)

More countries then began to join the treaty: Germany joined on 8 April 1983, followed by Norway on 24 January 1992, France on 2 July of the same year, Italy on 30 March 1993, Belgium on 21 November 2002, and Finland on 3 July 2004.

There was much to celebrate with the outcome. The Treaty of Cooperation in Astrophysics covers many matters of great importance:

- The Observatories would be Spanish and governed by Spanish legislation in addition to the stipulations in the Treaty.
- The IAC would be granted ownership of both Observatories.
- Spain would be awarded, free of charge, at least 20 per cent of the observing time on each of the telescopes and instruments installed at the Observatories (this observing time, under the supervision of the IAC, would be for the use of Spanish institutions and other collaborative centres of any nationality, with the addition of a further 5 per cent of observing time on each telescope assigned for carrying out cooperative programmes).
- Those institutions with instrumentation at the IAC Observatories would cooperate in the training of Spanish scientific and technical personnel, with grants from the owner countries, and programmes of scientific and technical cooperation would also be set up.

- Foreign institutions would have the right to the use, free of charge, of any land necessary for buildings to house their telescopes and instruments, any such telescopes and instruments remaining the property of their institutions of origin, even should these decide to withdraw from the Treaty.
- The property, and the administration and management of the services of the Observatories would be the responsibility of the IAC, which would charge for maintenance and use.

Just as petroleum is measured in barrels of quality crude, the high astronomical quality of the sky is a singular natural resource requiring its own unit of measure: the 'number of observing nights'. The more and better the telescopes, and the greater the astronomical quality of the sky and observatory infrastructures, the more prized is the observing time. So that those unversed in astronomy may comprehend the importance of the achievements of the Treaty, it is useful to note that it is normal in similar accords to be granted 5 (at most 10) per cent of the observing time, and that we had obtained 20+5 per cent. In other words, we had obtained a quarter of the total available observing time, which has been of fundamental importance in the development of Spanish astrophysics. We had gone from having no modern telescopes to availing ourselves of 25 per cent of the observing time on some of today's most advanced telescopes, located on an exceptional site. This wealth of high-quality observing time began to be used as a lever to bring about collaborations between young astrophysicists and the most distinguished authorities in the world. That is how we have reached such a high level in this branch of science in so short a time. As newcomers, we had been able to use our celestial capital wisely so that our successors could rise in status.

Attaining these heights was not easy. Even today, many ask how it could have been done. I believe there were two reasons for our success. First of all, our partners were fighting against the clock to hold on to the funding they needed to finish and set up their telescopes. Second, and more importantly, given the small number of Spanish astrophysicists at the time, our partners were sure that we would be unable to use such an abundance of observing time, in which case the excess time would revert to themselves. They were mistaken; in ten years their own statistics have shown that we were using up around 40 per cent of the observing time. In 1989 the Isaac Newton Group of Telescopes reported that 40 per cent of the Group's scientific production was owed to observations made with their telescopes by Spanish astrophysicists.

I must emphasize, particularly now that we are in an age of autonomies and secessionism in Spain, that in drafting the Treaty we did not fall into the

Fig. 10.4 Participants at a meeting of the Scientific Research Committee in 1985 at the entrance to the new headquarters of the IAC (then still under contruction). (Credit: IAC)

selfish trap of campaigning for 'Canarian astrophysics'. We negotiated these successful cooperation agreements for the benefit of Spain's entire astronomical community. And still today observing time on the telescopes at the IAC Observatories is divided equally, according to the same set of rules, for all Spanish astronomers, whether or not they are based in the Canaries.

The Treaty had an initial duration of 30 years, automatically renewable every ten years, unless Spain should decide to communicate to the other participating countries its withdrawal from the agreement two years prior to any of these periods. The partner countries may withdraw from the Treaty at the end of any of these periods by communicating their intentions to the Spanish government at least two years prior to any of these periods under the obligation of removing all their property from the Observatories.

As the thirtieth anniversary of the signing of the Treaty approached, the Danish Director of the Nordic Telescope made moves to persuade the signatory countries to pressure Spain to renegotiate the terms of the Treaty. His aim was to try and lower the 20+5 per cent of observing time 'toll' levied for access to the Canarian skies. Why, he reasoned, should the tenants of a building not unite to argue for lower rents? His argument was that Spanish astrophysics now ranked among the first in Europe, for which reason it no long had need of help in training scientific and technical staff, or so much observing time for itself. We had a struggle to counter this move, above all—strange as it may seem—with the authorities of our own Ministry, which seemed to have its head in the clouds. We managed to contain the protests of Denmark, Norway, and Sweden (Finland, part-owner of the Nordic Telescope, had not joined them). The Nordic countries did not after all leave, as might have been formally expected after they had lodged their protests.

Taking a broad view of what has been achieved, the advantages that have accrued for science and technology in general (and not only for Spain) are evident. Powerful telescopes from international consortia continue to arrive at the Canarian Observatories and will continue to do so while ground-based observations of the Universe continue to be carried out.

But we cannot rest on our laurels. New observing instruments that improve the performance of their predecessors continue to evolve and there is a permanent fight to choose the best sites for their emplacement. For that reason, the IAC maintains a very active team dedicated to measuring the parameters that characterize the astronomical quality of the atmosphere above high mountain sites. This strategy continues to bear excellent fruit. The astronomical promotion of the Canarian skies is a continuing hard-fought crusade.

11

Telescopes Versus Military Radars (1977–1978)

The previous chapter described how the Treaty of Cooperation in Astrophysics was finally signed. I now deal with a curious train of episodes that greatly impinged on that effort.

After years of effort, we were on the point of successfully ending the negotiations with scientific institutions from European countries actively designing the most advanced telescopes of the day. We were busy finalizing points in the draft agreements that we were anxious to sign off. At last, we were on the verge of endowing the Canary Islands with the most advanced astronomical instrumentation then available.

Then one morning I received a call from Modesto Fraile, Civil Governor of Tenerife. Without mincing his words, he announced, 'I regret to inform you that you will have to forget about Roque de los Muchachos as an observatory because the Ministry of Defence has decided to build an important radar system on the summit of La Palma.' All our dreams shattered in a single blow! But I was never one to surrender or take no for an answer. I asked the Governor if I could at least take measures to try and get this decision overturned. He replied sardonically that I could do whatever it entered my head to do with no opposition from him (that was already a triumph), but that 'it was Defence we were up against' and nothing would come of it. At the time the military were an all-powerful force and were still to be feared.

To be thus thwarted just as we could almost touch the marvellous telescopes that were about to crown the peaks of the Canary Islands!

Almost in desperation, I took decisive action without delay. I immediately telephoned the Chiefs of Staff. After an hour of being bounced from one extension to another, I finally managed to contact an official who knew

© The Author(s), under exclusive license to Springer Nature Switzerland AG 2021
F. Sánchez, *The Rise of Astrophysics in Modern Spain*, Astronomers' Universe,
https://doi.org/10.1007/978-3-030-66426-8_11

something about the topic. From him I learned, 'The western world being unprotected in an area of the North Atlantic, it was vital and urgent to remedy this gap in our defences by setting up a powerful radar system on the summit of La Palma'. NATO, he continued, and of course the Americans, felt that Spain was responsible for covering this flank in the West's defences. The Cold War was at its height, and the Regime needed to cosy up to the so-called western powers. Moreover, he informed me, that 'they' had already visited Roque de los Muchachos (using the very road that ICONA had just built), that 'that was the site', and that there was nothing more to be said on the matter.

Determined not to give up the fight, I countered as best I could by listing the many scientific benefits that the observatory about to be built on the crest would bring. I told him the advanced stage that had been reached with the international agreements that would be most advantageous for Spain. I failed to penetrate the stone wall.

I had a sudden flash of inspiration and put myself inside the military head of the chief of staff with whom I was speaking. Not for nothing was I the son of a military father and a reserve officer of the glorious Spanish army. I decided to bombard my interlocutor with the Motherland and History! But it had to be done respectfully: in those days one did not treat such matters lightly. I tactfully assumed a military style and terminology.

I began by accepting that the defence of the country was a top priority, and that, in cases of possible conflict between research and defence, the latter must prevail. I then informed the official that the Americans were building modern telescopes on Mauna Kea, on the highest point of the Hawaiian Islands. This was clear evidence they had found a formula to make science compatible with defence in Hawaii, in spite of the painful lesson of Pearl Harbor. Might we, then (that is to say, the Spanish army), not be held to international ridicule if we were to prevent, on the grounds of military defence, the building of European telescopes on La Palma? Ought we really be trying to out-pope the Pope? Our military, I said, should seek a military solution, as the Americans had done, that would save this major European project, a project through which we aimed to reach the highest scientific and technological level in astrophysics. I intoned melodramatically that the Motherland and History would judge harshly any failure to do so. I ended by urging them to take care not to look foolish in front of their international colleagues. He remained silent and said suddenly, 'Professor, could I call you back in a little while? I need to confer with my superiors.' The torpedo had struck home!

He called me back soon after to ask me if I could go to Madrid on the following day to discuss the matter in detail with them. I answered that of course

I would. I put down the phone, leaned back in my chair and took a deep breath. We had managed to gain entry to the 'Bunker' (as the upper echelons of the military were known during the Transition)!

I had to arm myself quickly to do battle. I convened my colleagues, at the time only a small handful of persons, so that we could discuss this grave situation together. In the early days of the IAC nobody knew anything about radar. The one closest to the topic was our electronics expert Sergio González, who immediately learned all he could about defence radar, for he would be the one accompanying me Madrid. And at the time there was, of course, no internet.

We took the first plane on the following morning and arrived at the Air Force Chiefs of Staff for our appointment. We entered the impressive neo-Escorial style building of the ministry that I had previously seen being finished at the Moncloa when I was at university. We were met by five or six high-ranking military officers. They spent much time courteously and precisely making us understand the imperious and urgent need to set up a powerful radar system on the very summit of Roque de los Muchachos. So powerful were these instruments, they explained, that they could fry any unlucky bird that happened to pass through their sweeping beams, thus rendering it utterly impossible for us to put our telescopes there. I in turn explained to them the advanced stage reached in the agreements, underlining the benefits that these would bring for our country and for Europe. I ended my peroration by repeating my patriotic appeal that had previously made such an impact. It was a very long, tense, but civil meeting in which we all agreed that Spain must both defend itself and produce competitive science and technology—all within an international framework, naturally.

I can well remember a certain Colonel Jack (he was Spanish in spite of his surname), who had just returned from the United States and seemed to be the best-informed concerning the blessed radars, coming round to half-accepting the possibility of coexistence. Perhaps it might be possible to put the radars on another island, or even on La Palma below the telescopes plus a smaller radar on the summit to cover the shadow produced by Roque de los Muchachos. I thought I saw in his face a sharp inner conflict as he debated with himself the possibility of their having after all to alter their original plans. They ended the meeting with a promise that they would study the options and send a team of experts to La Palma. They would then call us to inform us of the final decision.

In the spring of 1978 President Adolfo Suárez came, with a generous purse, on a campaigning tour of the Canary Islands. It was an electioneering visit that received great media coverage. He spent more than a week going from island to island, dispensing largesse, holding meetings, and conferring with corporations, businesses, and private individuals. He even wanted to include the University in his schedule. But instead of going to the University of La Laguna, which was the only one in the archipelago at the time, he summoned the Governing Council of the University to the headquarters of the Civil Governor of the Province of Santa Cruz de Tenerife. The universities were in turmoil during the Transition, and campuses were anything but tranquil. To inflame the situation further, the chosen venue was a pointed gesture to dot the i's and cross the t's, thus making it quite clear that the University was dependent on the State and should know its place in the broader scheme of things. University autonomy was still a distant dream (as explained in Chap. 9).

The Chancellor of La Laguna (then Professor Bethencourt) had prepared a careful discourse, praising the most important aspects of our University and dramatizing its most glaring deficiencies. Every effort needed to be made to squeeze as much money as possible out of the President in order to shore up the penurious economy of the University. His discourse then went on to mention the activities of the IAC, underlining the glorious future of astrophysics in the Canary Islands and the mainland if we could but succeed in signing the cooperation agreements being negotiated with Europe (as Vice-chancellor of Research, I had also participated in the drafting of the address).

When the Chancellor came to the requirements of the IAC for securing the arrival of large European telescopes at Roque de los Muchachos, and particularly to the building of a dorsal road for La Palma, Suárez interrupted him, saying that he had just come from that island, and that, when he had questioned the mayors specifically on whether they would prefer to spend 100 million pesetas (600,000 euros) on astrophysics or use that money to repair the poor roads connecting their towns, they had all opted for the latter. That was when I heard myself saying, 'Mister President, I am hardly surprised, given the way in which you put your question, that they gave you the answer you wanted to hear.' It was an unconsciously instinctive impulse, and my egregious faux pas was greeted in general silence. I soldiered on, in an effort to justify my temerity, and explained with clarity and passion to the President what had been happening. I stressed the importance of what our whole enterprise meant for Spanish science and technology. I described to him the advanced stage that had been reached in the international agreements that would enable Spain to host the most advanced and complete array of European telescopes. I also mentioned the hurdle of the radars and the American solution to a very similar problem in Hawaii.

When the Chancellor ended his discourse, the President gave his, and at the close of the meeting, Suárez shook hands with all the members of the

Governing Council. When he got to me, he took my hand and put a friendly arm around my shoulder. 'Don't worry, Professor,' he said, 'they're not going to take the European telescopes to Hawaii.' Years later, when he was no longer president, we met at a Prince of Asturias Awards ceremony and he still remembered my boldness.

The next meeting with the Air Force Chiefs of Staff was very different; things had changed now that General Gabeiras (then Secretary of State for Defence) had become aware of the situation. The military officers present, who had flown to La Palma looked as though they had come straight from the battle front, complete with bruises and cuts. Their helicopter, we were told, had made 'too complicated a landing' at Roque de los Muchachos, in heavy wind. They explained to us that they had now found a solution that would station the radars on the highest peak of the island of El Hierro, with possibly a very small one below Roque de los Muchachos Observatory. To this day, not a single radar has been erected on El Hierro, and the inhabitants of that island continue to fight against their ever being brought there.

The Treaty concerning Cooperation in Matters of Astrophysics was finally approved by the Council of Ministers at the joint request of the Ministries of the Presidency, Education and Science, and (naturally!) Defence. It was then

Fig. 11.1 President Adolfo Suárez greeting senior military officers and Canarian politicians during his 1978 visit to the islands. (Credit: *Diario de Avisos*)

Fig. 11.2 Placard in the outskirts of Valverde (El Hierro) protesting against the installation of military radars on the summit of Malpaso. The banner reads: 'No base, no radar. El Hierro for peace.' (Credit: Flickr/José Mesa)

countersigned by Congress and published in the Official State Bulletin. The signing took place at the Island Cabildo of La Palma on 26 May 1979 in the presence of ambassadors from the signatory countries and the Presidency of the Government Minister (then José Pedro Pérez-Llorca). It was the first time that Spain had signed such a high-level treaty in the Canaries. From that moment the doors of the Observatories of the Instituto de Astrofísica de Canarias were formally opened, and have remained open ever since, to the international scientific community, and the Observatories began to be populated with latest-generation telescopes and instruments.

Astrophysics had triumphed, as borne out by today's splendid panoramas at Teide Observatory and Roque de los Muchachos Observatory, where telescopes now flourish.

Fig. 11.4 Panoramic view of Roque de los Muchachos Observatory on La Palma. (Credit: Daniel López/IAC)

Fig. 11.3 Panoramic view of Teide Observatory today. (Credit: Daniel López/IAC)

12

Science Meets Politics: The Law of Astrophysics and Its Avatars (1975–2011)

By the mid-seventies we had made quite a lot of progress with regard to the juridico–administrative status of astrophysics in the Canary Islands. It was one of our key strategic objectives, but much still remained to be done. I now describe how the task was completed and the difficulties we still face today in maintaining this particular form of identity.

First of all, it was vital to prevent the newborn discipline of astrophysics in the Canaries from being swallowed up and lost in the centralizing whirlpool of CSIC or some other Madrid-based body. We achieved this by following a two-pronged strategy: (1) founding the University Institute of Astrophysics of the University of La Laguna, which incorporated Teide Observatory, and (2) simultaneously creating, through the Provincial Inter-island Association of Cabildos of the Province of Santa Cruz de Tenerife, the Astrophysics Institute of the Association of Cabildos, so that lands could be ceded for what was to become Roque de los Muchachos Observatory.

In 1975 a further step forward was taken towards administrative consolidation by unifying all astrophysical activity in the islands under a single entity, of which CSIC, a scientific body of national status, would be simply one member more participating in this major astrophysical project. All this formed part of an agreement of cooperation signed in Madrid, in the presence of the Minister for Education and Science, by the Chancellor of the University of La Laguna, the President of the Provincial Inter-island Association of Cabildos of the Province of Santa Cruz de Tenerife, and CSIC. This new entity was called the Instituto de Astrofísica de Canarias (IAC).

But the decisive step was finally taken in the eighties, when the Instituto de Astrofísica de Canarias was granted by law 'juridical personality with full

© The Author(s), under exclusive license to Springer Nature Switzerland AG 2021
F. Sánchez, *The Rise of Astrophysics in Modern Spain*, Astronomers' Universe,
https://doi.org/10.1007/978-3-030-66426-8_12

capacity to act'. The repercussions of this step on the growth of astrophysics were immediate.

The Institute's novel status even influenced subsequent legal norms applied to academic and scientific bodies throughout the country. But new ideas give rise to suspicion and envy, and the 'Astrophysics Law' has had its share of flak from way back in the eighties until the present day. The urge on the part of civil servants in any administration to unify and standardize, with their dedication to 'proper channels' and 'precedents', was, and continues to be, another permanent hindrance for the IAC. We shall also speak of this because it is a persistent obstacle set up by bureaucratized administrative structures that bog down and stifle progress in any country.

As mentioned earlier, in 1979, after years of hard negotiations, something of major importance had been achieved: several European countries had signed a multinational treaty for the IAC Observatories to host telescopes from all over the world and to safeguard the construction, use, and ownership of such advanced instrumentation. This treaty was ratified by the parliaments of the signatory countries and remains in force and operative today, and on this solid foundation rests the splendour of the Canarian Observatories. I emphasize this because it had a decisive influence on the passing of the Astrophysics Law.

After this, we understood that the time had come for us to take the definitive step in tackling the juridical formalization of the IAC that would give it genuine capabilities of self-management and governance. We needed these capabilities in order to exploit all the possibilities that were now open to us, capabilities that could not be granted by the University of La Laguna, CSIC, or the island cabildos.

It was plain to us that we needed a bold and wily legal expert able to think outside the box. Luckily, a young legal academic, Professor Gaspar Ariño, who had just arrived from the United States on a Fullbright grant, had recently been incorporated into the Department of Administrative Law at the University of La Laguna. He was an expert in the formal relations between academic institutions and research centres, so I wasted no time in seeking his help and counsel.

I explained to him what the IAC was and where we wanted to go. I spoke to him of our longed-for goals and possibilities, and I presented him with the following challenge: to design an entity capable of endowing the IAC with the necessary legal capacities, in spite of its being a Spanish institution, in order to make it genuinely competitive on an international level. That entity would need to possess the sufficient juridico–administrative capacity to do what we wanted it to do, which was astrophysical research, but also to be capable of assimilating and transferring the advanced technology associated with big scientific projects. We likewise wanted to train scientists and technologists, and spread scientific knowledge through astronomical outreach. I saw at once that he warmed to the idea, and he climbed on board immediately. He was hungry with an appetite for the challenge.

We obtained an economic grant from the nascent Autonomous Community of the Canary Islands to assign the project to the Department of Administrative Law of the University of La Laguna. Professor Ariño soon prepared a draft outlining the subject but he discovered that, under Spanish legal regulations in force at the time, the project would be impossible. He advised us that a new law, specifically dedicated to the IAC, would be necessary—not, in his view, a viable proposition given the extreme weakness in which the UCD government found itself at that time (the early eighties). On 28 January 1981 Adolfo Suárez resigned as President and as leader of the UCD; shortly after, on 23 February, there occurred a failed coup d'état, right in the middle of the vote for the investiture of Leopoldo Calvo-Sotelo; the economic crisis and the disintegration of the UCD led to its resounding electoral defeat in October 1982, followed by the dissolution of the party in February 1983. Professor Ariño had been perfectly correct in his frank assessment: 'the oven wasn't yet ready to bake the buns'.

Nevertheless, the fact remained that we now had a juridically very well structured and cogently argued, but novel, piece of planned legislation. It rested on new administrative solutions for modern research centres, but it strayed from the well-trodden path, an approach that always raises eyebrows ('better the devil you know'), so it behoved us to proceed warily. It all looked fine on paper: it marshalled the activities and means of all the institutions with an interest in astrophysics, unifying them all in a single entity with a 'juridical personality with full capacity to act' in managing and governing its own affairs (a 'public management consortium').

What to do next, tuck it away in a drawer in the hope of better times? Once again, we had to push for the impossible and urge a government fearful of introducing new legislation to regard this as a special case. For the government even to consider doing this it needed guarantees of the necessary parliamentary support, so we set about explaining everything to and convincing the main political parties by speaking to the highest-profile Canarian representatives in each party. We managed to convince them of the importance and urgency of the matter and the pressing need to pass the 'Astrophysics Law'. The time was right at that particular political juncture because the UCD needed a piece of legislation that would have the majority support of Parliament time was running out. At last, leaders of all the parties went to the

Moncloa to talk to the President and guarantee the support of their respective parties. After all, it was only an innocuous question of science, devoid of party connotations. I imagine that the President agreed because he needed to carry the parliament with him in a show of unanimity.

The first step was taken on 21 September when the PSOE presented a proposal to 'provide the Instituto de Astrofísica de Canarias with scientific, technical, and administrative personnel' according to the requirements set out in the Treaty concerning Cooperation in Matters of Astrophysics Treaty.

Once assured of the support of the main parties, the government approved the bill at the Council of Ministers, and on 30 April 1982 the Royal Decree-Law 7/1982 was proclaimed, which stated, 'The Instituto de Astrofísica de Canarias is hereby created and its juridical regime established'. The law was published in the Official State Bulletin of 5 May 1982 and was presented to Parliament by the government, the decree-law procedure being justified by the urgency of the matter. It was debated by Congress in May and was defended by Mrs. Pelayo Duque on behalf of the Party of Democratic Action, Mr. Padrón Delgado for the Socialist Parliamentary Group, and Mr. Fernández Rodríguez for the Centrist Parliamentary Group. It is well worth reading, in *Diario de Sesiones del Congreso*, these Canarian politicians' clear arguments and enthusiasm for the IAC. The decree-law was finally validated by Congress on 11 May 1982, with 262 votes in favour and two abstentions.

This law constituted a turning point, not only for Spanish astronomy but also for subsequent legislation concerning Spanish science. It was the first time that the concept of a public consortium had been applied to a scientific body. The IAC had been granted a considerable degree of specific autonomy, placing it on a par with 'autonomous bodies of a commercial, industrial, financial, or analogous character', as identified in the law. Likewise, it could now hire its own personnel and that of associated administrations, together with that of 'other private or public entities with which the Institute agrees administrative or civil contracts, or agreements of cooperation'. These concepts would later be used in the new economic regime of the universities. All astronomical activity in the Canaries was encompassed in this law: 'The Institute of Astrophysics of CSIC and the University Institute of Astrophysics of the University of La Laguna are attached to the Instituto de Astrofísica de Canarias'.

The IAC was also firmly granted 'juridical personality and the capacity to act', among other things, 'so that Spain may properly meet its international obligations', a reference to the Treaty of Cooperation in Astrophysics. The law specifies that the Instituto de Astrofísica de Canarias has juridical personality and the capacity to act in order to fulfil the following ends:

(a) To carry out and promote any kind of astrophysical research or other research related thereto.

(b) To spread astronomical knowledge, collaborate in university teaching specializing in astronomy and astrophysics in the university district of La Laguna, and train scientific and technical personnel in all fields related to astrophysics.

(c) Administer the existing centres, observatories, astronomical and facilities, and those to be created or incorporated under its administration in the future, together with any entities dependent on its services.

(d) To foster relations with the national and international scientific communities.

Those of us working at the *Astrofísico* in the spring of 1982 celebrated our newly promulgated law in grand style, with a dinner and much revelry at a very popular restaurant near the runway of Los Rodeos airport. We were rather boisterous and, not to put too fine a point on it, we rather overdid it. Such were our antics and horseplay that the owner of the restaurant told us to settle our bill and never darken her door again.

The new IAC was established at a high-level meeting of its Governing Council, attended by the Minister of the Presidency of the Government, the President of the Canarian Council (this was the title of the Canarian government in that pre-autonomous time), the Chancellor of the University of La Laguna, the President of CSIC, and the Director of the Institute (an ex oficio member and Secretary of the Governing Council). The Council met at the Moncloa, the headquarters of the Presidency of the Government, on 26 July 1982 and began by appointing the Minister of the Presidency of the Government as Chair and Professor Francisco Sánchez as Director of the IAC.

The new status granted to the Instituto de Astrofísica de Canarias by this new law was ill received by the National Geographical Institute, to which the National Commission of Astronomy, the Spanish part of the German Calar Alto Observatory (Almeria), and the National Observatory (as the Astronomical Observatory of Madrid was then called) were all answerable (each of them also had designs on controlling the Canarian Observatories). CSIC was even less favourable for the same reasons because it had been given to understand that the IAC would be just one CSIC-controlled institution more. It was unthinkable that a minor entity like the IAC should have prerogatives and privileges exceeding those of the all-powerful CSIC, even if only in certain areas. These feelings of affront, together with the rising star of the IAC within and beyond our frontiers, generated grudges in many minds against the Institute that have harmed all involved. We have always been conscious of the permanent

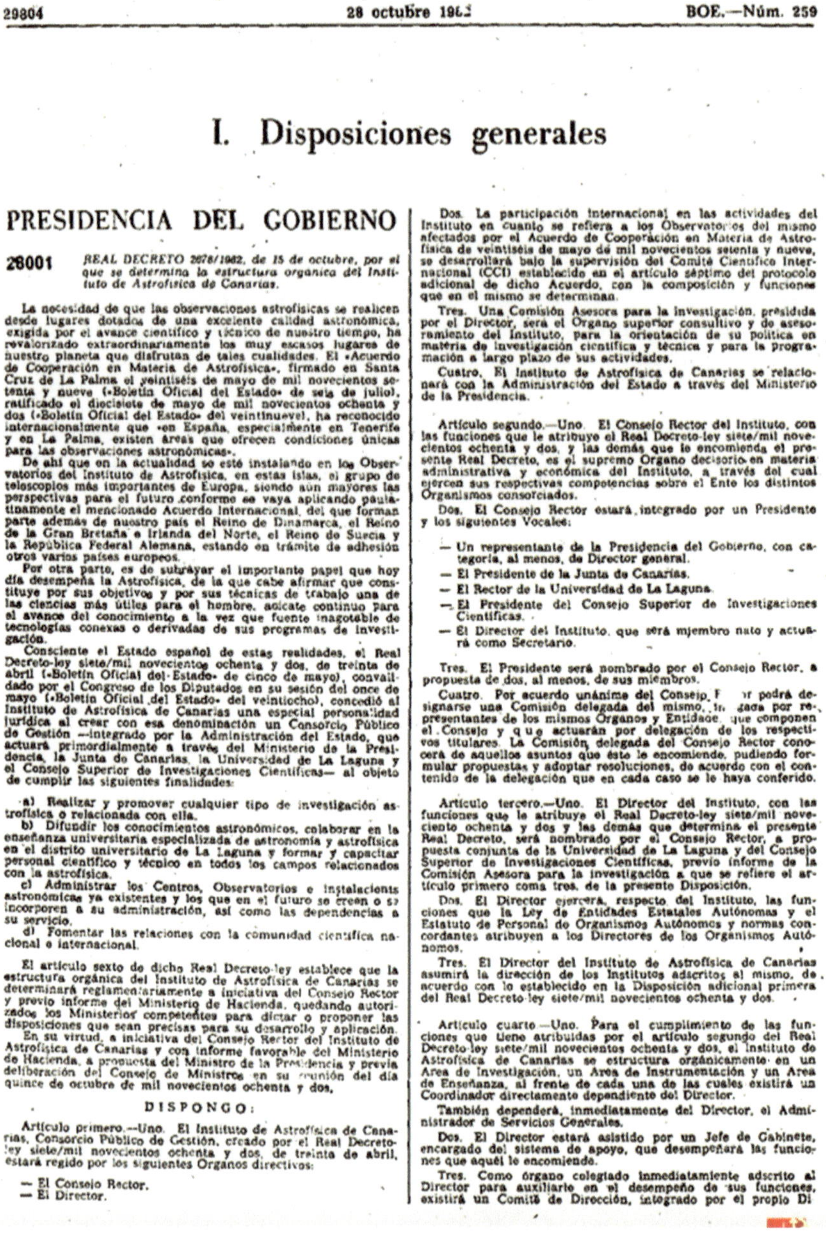

I. Disposiciones generales

PRESIDENCIA DEL GOBIERNO

28001 *REAL DECRETO 2678/1982, de 15 de octubre, por el que se determina la estructura orgánica del Instituto de Astrofísica de Canarias.*

La necesidad de que las observaciones astrofísicas se realicen desde lugares dotados de una excelente calidad astronómica, exigida por el avance científico y técnico de nuestro tiempo, ha revalorizado extraordinariamente los muy escasos lugares de nuestro planeta que disfrutan de tales cualidades. El «Acuerdo de Cooperación en Materia de Astrofísica», firmado en Santa Cruz de La Palma el veintiséis de mayo de mil novecientos setenta y nueve («Boletín Oficial del Estado» de seis de julio), ratificado el dieciséis de mayo de mil novecientos ochenta y dos («Boletín Oficial del Estado» del veintinueve), ha reconocido internacionalmente que «en España, especialmente en Tenerife y en La Palma, existen áreas que ofrecen condiciones únicas para las observaciones astronómicas».

De ahí que en la actualidad se esté instalando en los Observatorios del Instituto de Astrofísica, en estas islas, el grupo de telescopios más importantes de Europa, siendo aún mayores las perspectivas para el futuro conforme se vaya aplicando paulatinamente el mencionado Acuerdo Internacional del que forman parte además de nuestro país el Reino de Dinamarca, el Reino de la Gran Bretaña e Irlanda del Norte, el Reino de Suecia y la República Federal Alemana, estando en trámite de adhesión otros varios países europeos.

Por otra parte, es de subrayar el importante papel que hoy día desempeña la Astrofísica, de la que cabe afirmar que constituye por sus objetivos y por sus técnicas de trabajo una de las ciencias más útiles para el hombre, aúicate continuo para el avance del conocimiento a la vez que fuente inagotable de tecnologías conexas o derivadas de sus programas de investigación.

Consciente el Estado español de estas realidades, el Real Decreto-ley siete/mil novecientos ochenta y dos, de treinta de abril («Boletín Oficial del Estado» de cinco de mayo), convalidado por el Congreso de los Diputados en su sesión del once de mayo («Boletín Oficial del Estado» del veintiocho), concedió al Instituto de Astrofísica de Canarias una especial personalidad jurídica al crear con esa denominación un Consorcio Público de Gestión —integrado por la Administración del Estado, que actuará primordialmente a través del Ministerio de la Presidencia, la Junta de Canarias, la Universidad de La Laguna y el Consejo Superior de Investigaciones Científicas— al objeto de cumplir las siguientes finalidades:

a) Realizar y promover cualquier tipo de investigación astrofísica o relacionada con ella.

b) Difundir los conocimientos astronómicos, colaborar en la enseñanza universitaria especializada de astronomía y astrofísica en el distrito universitario de La Laguna y formar y capacitar personal científico y técnico en todos los campos relacionados con la astrofísica.

c) Administrar los Centros, Observatorios e instalaciones astronómicas ya existentes y los que en el futuro se creen o se incorporen a su administración, así como las dependencias a su servicio.

d) Fomentar las relaciones con la comunidad científica nacional e internacional.

El artículo sexto de dicho Real Decreto-ley establece que la estructura orgánica del Instituto de Astrofísica de Canarias se determinará reglamentariamente a iniciativa del Consejo Rector y previo informe del Ministerio de Hacienda, quedando autorizados los Ministerios competentes para dictar o proponer las disposiciones que sean precisas para su desarrollo y aplicación.

En su virtud, a iniciativa del Consejo Rector del Instituto de Astrofísica de Canarias y con informe favorable del Ministerio de Hacienda, a propuesta del Ministro de la Presidencia y previa deliberación del Consejo de Ministros en su reunión del día quince de octubre de mil novecientos ochenta y dos,

DISPONGO:

Artículo primero.—Uno. El Instituto de Astrofísica de Canarias, Consorcio Público de Gestión, creado por el Real Decreto-ley siete/mil novecientos ochenta y dos, de treinta de abril, estará regido por los siguientes Órganos directivos:

— El Consejo Rector.
— El Director.

Dos. La participación internacional en las actividades del Instituto en cuanto se refiera a los Observatorios del mismo afectados por el Acuerdo de Cooperación en Materia de Astrofísica de veintiséis de mayo de mil novecientos setenta y nueve, se desarrollará bajo la supervisión del Comité Científico Internacional (CCI) establecido en el artículo séptimo del protocolo adicional de dicho Acuerdo, con la composición y funciones que en el mismo se determinan.

Tres. Una Comisión Asesora para la Investigación, presidida por el Director, será el Órgano superior consultivo y de asesoramiento del Instituto, para la orientación de su política en materia de investigación científica y técnica y para la programación a largo plazo de sus actividades.

Cuatro. El Instituto de Astrofísica de Canarias se relacionará con la Administración del Estado a través del Ministerio de la Presidencia.

Artículo segundo.—Uno. El Consejo Rector del Instituto, con las funciones que le atribuye el Real Decreto-ley siete/mil novecientos ochenta y dos, y las demás que le encomienda el presente Real Decreto, es el supremo Órgano decisorio en materia administrativa y económica del Instituto, a través del cual ejercen sus respectivas competencias sobre el Ente los distintos Organismos consorciados.

Dos. El Consejo Rector estará integrado por un Presidente y los siguientes Vocales:

— Un representante de la Presidencia del Gobierno, con categoría, al menos, de Director general.
— El Presidente de la Junta de Canarias.
— El Rector de la Universidad de La Laguna.
— El Presidente del Consejo Superior de Investigaciones Científicas.
— El Director del Instituto, que será miembro nato y actuará como Secretario.

Tres. El Presidente será nombrado por el Consejo Rector, a propuesta de, al menos, de sus miembros.

Cuatro. Por acuerdo unánime del Consejo Rector podrá designarse una Comisión delegada del mismo, integrada por representantes de los mismos Órganos y Entidades que componen el Consejo y que actuarán por delegación de los respectivos titulares. La Comisión delegada del Consejo Rector conocerá de aquellos asuntos que éste le encomiende, pudiendo formular propuestas y adoptar resoluciones, de acuerdo con el contenido de la delegación que en cada caso se le haya conferido.

Artículo tercero.—Uno. El Director del Instituto, con las funciones que le atribuye el Real Decreto-ley siete/mil noveciento ochenta y dos y las demás que atribuye el presente Real Decreto, será nombrado por el Consejo Rector, a propuesta conjunta de la Universidad de La Laguna y el Consejo Superior de Investigaciones Científicas, previo informe de la Comisión Asesora para la Investigación a que se refiere el artículo primero, coma tres, de la presente Disposición.

Dos. El Director ejercerá, respecto del Instituto, las funciones que la Ley de Entidades Estatales Autónomas y el Estatuto de Personal de Organismos Autónomos y normas concordantes atribuyen a los Directores de los Organismos Autónomos.

Tres. El Director del Instituto de Astrofísica de Canarias asumirá la dirección de los Institutos adscritos al mismo, de acuerdo con lo establecido en la Disposición adicional primera del Real Decreto-ley siete/mil novecientos ochenta y dos.

Artículo cuarto.—Uno. Para el cumplimiento de las funciones que tiene atribuidas por el artículo segundo del Real Decreto-ley siete/mil novecientos ochenta y dos, el Instituto de Astrofísica de Canarias se estructura orgánicamente en un Área de Investigación, un Área de Instrumentación y un Área de Enseñanza, al frente de cada una de las cuales existirá un Coordinador directamente dependiente del Director.

También dependerá, inmediatamente del Director, el Administrador de Servicios Generales.

Dos. El Director estará asistido por un Jefe de Gabinete, encargado del sistema de apoyo, que desempeñará las funciones que aquél le encomiende.

Tres. Como órgano colegiado inmediatamente adscrito al Director para auxiliarle en el desempeño de sus funciones, existirá un Comité de Dirección, integrado por el propio Di

Fig. 12.1 Announcement in the *Boletín Oficial de Estado* concerning the creation of the Instituto de Astrofísica de Canarias by a Royal Decree of 15 October 1982. (Credit: Ministry of the Presidency, Government of Spain)

aspiration of CSIC to have the IAC under its wing. Such are the human foibles arising from competition, envy, and the lust for power.

The development of the IAC was initiated by the Royal Decree of 15 October 1982, which specified its competencies, functions, and organic structure. Noteworthy for the time was the contracting by the Consortium of non-civil servants. Furthermore, an employment grade for researchers, equivalent to that within CSIC, was created. With these arrangements the IAC became very free and agile, with a structure closely resembling that of a business with its own administrative body (the Governing Council) and a kind of deputy (the Director). It was inevitable that this should cause alarm and unease in the archaic scientific echelons of the Administration.

President Calvo-Sotelo was obliged to bring forward the elections, which were held on 28 October 1982. Felipe González's PSOE won by a large majority. Javier Moscoso occupied the Ministry of the Presidency of the Government and was therefore also the President of the IAC. It must be admitted that being answerable directly to the Presidency of the Government had been a great advantage and allowed the swift development of the Institute and its Observatories. I have fond memories of how well attended and treated we were there during Moscoso's mandate, especially by his Office Director, Pilar de la Torre, and by the head of the Budget Office of the Presidency of the Government, Julián Martín Arias—names not to be forgotten.

Our story now takes us to 1985 and the splendid inaugurations of the Institute's Observatories. This event put the IAC in the limelight, for both good and ill, and raised yet further hackles. It turned out that Professor Maravall, then Minister for Education, Universities, and Research, had envious designs, as he himself would tell me years later, with the result that, in a new Law of Science then being drafted, it was decided to bring the IAC within his ministerial fold.

The draft of this law reached the Council of Ministers on three occasions, with the intention of detaching the IAC from the Ministry of the Presidency and placing it under the wing of Education, Universities, and Research, thereby reducing it to one CSIC institute more and destroying its status as an autonomous consortium. Minister Moscoso had on a number of occasions managed to foil the manoeuvre. We were relieved when the law was sent to Parliament free from this amendment and voted through in Congress. But when the bill was sent for the Senate's approval, it was the socialists themselves (surprise, surprise) who introduced an amendment to give Maravall what he coveted.

We found ourselves suddenly at the cliff's edge, but we did not yield. We fought with astuteness and courage, winning over the public and Canarian politicians of all parties to our side. The Canarian socialists were key players;

they knew us well and had been joint protagonists in the formidable advance made by the IAC because of its powerful juridico–administrative status. There were dramatic moments in which the Canarian news media and citizenry took up cudgels. There was an air of indignation. Stevedores at the port of Santa Cruz even offered to go on strike to 'save *el Astrofísico*'.

On 20 July 1986, the local newspaper *El Día* published a large and expressive cartoon with Minister Maravall driving a steamroller bearing the name 'Ministry of Education' over Francisco Sánchez, who has 'Instituto de Astrofísica de Canarias' written on his forehead, while the UCD deputy Luis Mardones struggles vainly to halt the machine, and the socialist Canarian President, Jerónimo Saavedra, joins his hands in prayer for Sánchez and the coming elections (repeated elsewhere in the newspaper). In another corner the minister Moscoso is seen launching a half volley with the ball of the 'accords and agreements' to Maravall. At the foot of the cartoon is written: 'The Instituto de Astrofísica is carted off for dead'.

It was a hard-fought battle. Moscoso and Saavedra gave it up for lost and withdrew. But neither I nor Dr. Alberto de Armas, a Canarian political historian who was then a PSOE senator, gave up the fight. I remember him with gratitude and great affection. He had an apartment in the Edificio España in Madrid, where he would receive me and we would chat. He finally managed

Fig. 12.2 A satirical cartoon published in the Tenerife daily newspaper *El Día* on 20 March 1986 concerning the Science Law and its consequences for the IAC inspired by the attempt by Minister Maravall to crush the 'IAC Consortium', thereby depriving the IAC of its juridical personality and capacity to operate. (Credit: *El Día*)

to get Maravall to agree to a pact in which he would agree to certain amendments to his proposal concerning the Institute. We finally managed, then, to stave off the demise of the IAC, which would after all retain its status as a consortium, although it would be wounded at its most sensitive point, its personnel. It was definitively established that the Ministry of Education and Science would oversee and manage this aspect. That is why the eleventh additional disposition of the Law concerning the Promotion and General Coordination of Scientific and Technical Research is dedicated entirely to the IAC. It is worth highlighting two key points of this added clause. First, that the Minister for Education and Science would chair its Governing Council. Second, the IAC's own staff would consist solely of non-civil servants dedicated to functions other than research, and the Grades of Astrophysicist and Adjunct Astrophysicist established by Royal Decree 2678/1982 of 15 October would be extinguished. Research personnel, then, would be provided by bodies associated with the IAC; that is to say, by CSIC, the University of La Laguna, and the Canarian Government. The IAC was thus bound hand and foot, and found itself in the extraordinary situation of being a research centre prohibited from hiring its own staff to carry out research. The Law of Science, that had so changed and generally benefited the system of Spanish research and development had undermined the very law that founded the IAC, thereby creating the first dysfunctions within the Consortium, above all in its staffing structure. These defects have taken years to set right, and even then have only been partially remedied.

The day after the new text of the Law of Science was approved, Minister Maravall called me personally; I had only ever previously seen him at the inaugural acts of the summer of 1985. I was greatly surprised, even more so when he began by politely apologizing—very courteous of him—for what he had done and for the way in which it had been done. He pleaded with me to understand why all this had occurred. 'The Minister of Science,' he explained, 'could not be put in the intolerable position, when asked about the IAC on foreign visits, of not being able to say anything.' He continued by affirming, 'That is why it was essential that so important a research centre should be directly answerable to my ministry'. He told me that he was willing to come immediately to the Canaries to talk it over with me, to which I replied that I was prepared to go and see him that very week and place my directorship at his disposal. He made an appointment for me to meet him in his office a few days later.

My witness at the meeting was Professor Teodoro Roca, recently appointed to the Chair of Astrophysics at the University of La Laguna. The minister was accompanied by Professor Juan Rojo, Secretary of State for Science and

Technology. The four of us were in discussion for the whole morning, all incoming calls being held. I began by stating my position clearly. I told him bluntly I knew that there were rumours being circulated in CSIC that the IAC was the fiefdom of Paco Sánchez, that its research was a disaster, and that its Spanish character was being put at risk, for which reason it was necessary to bring it back into the fold of CSIC and get rid of Paco Sánchez. I told them in all seriousness that they had an obligation to investigate properly whether any of these rumours were true, and that, were they found to be untrue, not to waste time in fixing what was running smoothly when they faced the uphill task of putting right the many things that genuinely needed fixing in Spanish scientific research. I ended by saying that of course my directorship of the IAC lay in their hands.

Juan Rojo, a renowned Professor Physics of the Complutense University, paid us several visits in the months that followed, even spending whole nights at the telescopes to confer with our astrophysicists. We passed his inspection with flying colours! They also came to realize that we had created—*they* had created—a serious problem by deleting our research staff with the stroke of a pen. We had to join efforts to 'right such an egregious wrong' with the Directorship General of Personnel Expenditure of the Ministry of Taxation (then located in a street just off Puerta del Sol—I remember it well from the numerous times I had visited it). It was necessary to adjust the decree on the basis of which the law governing the IAC was developed. The problem was finally eased by introducing a sound 'organic workforce' on the basis of funding for 'organic positions', a kind of opening that carries an associated stipend, to be filled through competition by serving civil servants who met the necessary requirements. The posts on offer were sufficiently desirable to attract researchers from the University and CSIC. We were thus able to attract very good people and re-employ our astrophysicists from the former Astrophysicist Grade.

But the avatars born of the far-sighted Law of the IAC continued to encounter problems. It would weary the reader for me to list meticulously all the obstacles that were laid in our path. Innovators and freethinkers have always had a hard time in Spain. Nothing brings greater calm to a civil servant tasked with drafting a regulation than the existence of a precedent. And it is civil servants who write the decrees that become law. What is certain is that the Administration, in its eagerness to impose uniformity, has gradually tightened its grasp on the IAC. This exaggerated hankering for uniformity is a general problem with which the Administration has burdened our research efforts, and those of other research bodies. A dull uniformity has restricted the flexibility needed to tackle the specifics of individual problems. We nevertheless have the satisfaction of witnessing how some of our 'advantageous innovations' have found their way into other organisms.

I outline here some of the restrictions placed on the IAC's ability to manage its own affairs efficiently. These obstacles were gradually introduced through successive budgetary laws, but in a particularly abrupt fashion through Law 14/2000 of 29 December concerning fiscal, administrative, and social order measures, and Law 14/2011 of 1 June concerning Science, Technology, and Innovation.

The Law concerning Science, Technology, and Innovation sought definitively to level out the differences in the regulations governing the Instituto de Astrofísica Public Consortium (IAC) by means of its lengthy twenty-seventh additional disposition, dedicated to the juridical regime of the IAC. This controversial law was introduced by Cristina Garmendia, Minister of Science, Technology, and Innovation, and her Secretary of State, Felipe Pétriz, when Mr. Zapatero was President. The most salient aspects of the disposition were that:

- The IAC was to become simply another Public Research Organism of the General State Administration.
- It would be left to the Statutes of the IAC, which must be approved by its Governing Council, to 'determine its organic, functional, and financial regime', always in accordance with 'what is established for public research organisms of the State Administration'.
- The mandate, explicitly stated in our Foundational Law, that the 'Governing Council shall be constituted by the Minister for Education and Science, who shall act as Chair, a voting representative of the State Administration, who shall be named by the Ministry of the Presidency, …' was to disappear, the determination of the composition of the Governing Council and all else that was important in making the IAC different being decided according to the Statutes.
- The regulations governing the Consortium, which formerly held the status of parliamentary law, were to be reduced to the level of juridico–administrative regulations, to be henceforth modified at the whim of the Governing Council.

Needless to say, during the drafting of this new law we mobilized and fought to ensure a different outcome. We did not stand a chance. Not even with the involvement of Parliament or the Senate was it possible to leave things as they were before, in spite of the best efforts of Juan Ruiz, then responsible for science, technology, innovation, and the information society on behalf of the Canarian Government. Neither could we get the new law to stipulate, as we would have wished, that the Director of the IAC should be elected by an independent international committee of astrophysicists of repute. Since I wanted to retire with an easy mind, I became obsessed with this matter. It simply could not be, as is habitual in large research centres in Spain, that the IAC's director should be appointed by the minister of the day according to party interests.

The Statutes needed to be drawn up, and we got down to work. We did what we always do at the IAC when confronted with a serious problem: we set up a working group to assist the IAC Director's office in its deliberations. It was a wide-ranging group made up of the Board of Governors and the most distinguished members of the IAC. We met regularly every week to investigate how best to influence the delicate and ever-changing process of drafting the Statutes. The Statutes were being drafted by the Ministry of Science, Technology, and Innovation, but by law they had to be approved by the Canarian Government. I recall that, in the letter of appointment to the working group, to give them encouragement, I wrote, 'Of course, it will not be easy. But, as on other occasions, we shall prevail by combining our grey matter, and acting with intelligence and decision'.

Although we did not achieve all that we had wished, we did make a number of gains. And, of course, we won on the issue of the Director being selected from candidates worldwide by an independent international scientific committee. We had brought off quite a coup. It was an epic battle and took two legislatures for us to achieve something so common-sensical. I was forced to delay my retirement until I was 77 years old, but the IAC had finally caused a new route to be taken in the regulation of public research organisms in Spain. This was to be my last battle. I am content that under this new procedure Professor Rafael Rebolo—a renowned astrophysicist and efficient manager— is now at the helm of the IAC.

I imagine that the politicians and senior civil servants of various state administrations, and the Autonomous Community of the Canary Islands were very happy to see the last this obstinate, stubborn, insistent, and demanding person that they added to the Statutes the following additional clause: 'The first Director of the IAC Don Francisco Sánchez Martínez, in recognition of his extraordinary labours in building astrophysics in Spain, shall be given the life-long, honorific title of Founding Director of the IAC'.

Fig. 12.3 Staff gather on the lawn of IAC headquarters in 2013 to bid farewell to Francisco Sánchez. (Credit: Miguel Briganty/IAC)

13

Royalty and Heads of State Above the Clouds (1985)

From the summer of 1985, thanks to *¡Hola!* ('Hello!') and other celebrity gossip magazines, people began to be aware of the existence of the Canarian Observatories and the Instituto de Astrofísica de Canarias when we were given coverage by television and communications media worldwide. Photographs of queens and princesses with their skirts wafting in the trade winds caught the media's fancy. How could it miss the opportunity to catch so much royalty together in the Canaries, with the added curiosity that they had gathered for such an exotic event as the inaugurations of some institute of astrophysics and its observatories? What was certainly a most serious matter was bringing together so many monarchs, heads of state, ministers, and leading scientists from all over the world for an 'astronomical' event. It was quite out of the ordinary and unforgettable, a rare event for any part of the world, and it caused quite a stir.

In June 1985, for almost a week the entire Spanish royal family, the King and Queen of Sweden, the King and Queen of Denmark, the King and Queen of The Netherlands, the President of the Federal Republic of Germany, the President of the Republic of Ireland, the Duke and Duchess of Gloucester, in representation of Queen Elizabeth II of the United Kingdom, along with ministers and sundry European authorities, and the international scientific community headed by five Nobel laureates descended on the islands of Tenerife and La Palma.

Such a cluster of celebrities in a remote Atlantic archipelago, gathered specifically to celebrate a scientific—rather than a military, political, or sporting—event was an altogether singular happening of a kind never before seen in Spain. It served first of all to help the Canary islanders discover that their

© The Author(s), under exclusive license to Springer Nature Switzerland AG 2021
F. Sánchez, *The Rise of Astrophysics in Modern Spain*, Astronomers' Universe,
https://doi.org/10.1007/978-3-030-66426-8_13

archipelago possessed something unique and very important in which they could take pride. From that moment on, *el Astrofísico* began to acquire recognition and popularity among our people, and Spaniards began to see in the Canaries something more than just a warm climate. The IAC had certainly given an extra unexpected and long-lasting added attraction to the Fortunate Isles.

Returning to the summer of '85, this magnificent display was not brought about by sheer chance. Everything was fraught with difficulty, starting with the challenge of getting so many VIPs to fit a purely scientific event into their crowded schedules. There were many complications. The inaugurations themselves had to be organized, along with the concomitant entertainments on both islands. And then there was the problem of transporting so many VIPs simultaneously. Extreme care had to be taken regarding security, logistics, protocol, accommodation, catering, and so forth. It was a tough challenge and a real headache for those charged with organizing it all, and everything needed to be carried out to perfection. It seems miraculous now that it all should have gone like clockwork and without a single hitch.

The inaugurations were the cherry on a rich cake that had been many years in the baking. To reiterate the ingredients: we had first of all to verify, with the scantest of means, that the skies above the Canarian summits met the most stringent requirements for observational astrophysics; then it was necessary to encourage and persuade foreign scientific institutions to cast off old prejudices and bring their modern telescopes here, and at the same time convince our politicians here that the Canarian skies were a valuable public 'natural resource' that could be tapped for science, technology, and culture, and even tourism.

A couple of amusing incidents served as light relief from all the solemnity of these inaugural events. When we began to think about what to do to leave a visible and lasting memorial of the occasion, we came up with the idea of a symbolic sculpture. And what better than a great spiral of the sort produced by the consummate Canarian artist Martín Chirino, cofounder in 1957 of the Spanish *El Paso* abstract movement? I spoke to him about it and he was delighted. He told me he would set to work on the concept immediately. We obtained money from the Ministry of the Presidency of the Government, to which the IAC was answerable at the time, and which was covering the costs of the inaugurations. But the President of the Canarian Government had found out about the plans and, without informing us, commissioned another Canarian artist, César Manrique, to do the sculpture. It was evident that Manrique was a friend of the President. Martín was naturally furious at this turn of events and did not speak to me for years, as though I had been to blame. César, on the contrary, became a friend.

But the matter did not rest there. As was the norm, I informed the International Scientific Committee (CCI) of the progress of events and I told them about the sculpture. Quite unexpectedly, when they examined Manrique's design there was uproar and general protest, fanned by our Scandinavian colleagues. They interpreted the sculpture as a phallus with Nazi connotations and were incensed. Under no circumstances, they declared, could the sculpture be given a place at Roque de los Muchachos Observatory. It turned out that certain members of the CCI had lived through the German occupation of their countries and were reviled by such symbolism. A solution was found that would avoid causing further offence. The CCI agreed to present to the authorities the notion that this sculpture gave rise to a serious scientific problem. Here is an excerpt from the agreement: 'The shape, size, and material of the proposed sculpture could produce serious perturbations in laminar air flow, provoking turbulence in the Observatory that would harm the excellent astronomical quality of its atmosphere'. In the light of such scientific considerations, the President of the Canarian Government bowed to the need to seek another site. The sculpture was finally relocated lower down the mountain near the road leading to the summit, and there it stands today, its sharp tip jutting upwards at the sky, a landmark for the cameras of tourists.

A question of deeper concern was security. The logistics of transporting so many heads of state to the Observatories almost simultaneously gave us many headaches. For a start, for reasons of security and to ensure smooth transport manoeuvres to avoid long delays we had to build not one but various heliports, both in Puerto de la Cruz (Tenerife) and at the two Observatories. And sufficient time had to be allocated for everything to be rehearsed beforehand. After what had happened to the helicopter carrying the military officers who wanted to install radars at Roque de los Muchachos, we were uneasy about landing helicopters in these terrains, so a trial flight was made, using the same type of helicopter that would be used at the inaugurations. The passenger list comprised the Head of Protocol of the Moncloa, the Head of Security of the Zarzuela, the Civil Governor of Tenerife, the Director of the IAC, and the General Chief of Staff of the Air Force in the Canary Islands, this last choosing to sit in the cockpit with the pilots. It was a military aircraft and we all wore interconnected headphones. We took off from Tenerife without incident, but when we were over the sea en route to La Palma, we overheard the commander of the helicopter say nervously, 'Don't touch that, General, we're descending'. To a man, we immediately lunged towards the cockpit and pulled the meddling general out. Just a few years earlier, such a daring act would have been unthinkable: military officers, especially high-ranking ones, were never wrong, and no civilian would ever have dreamed of contradicting them.

The programme of events began on 27 June, on the eve of the inaugurations, with an academic address at the auditorium of the University of La Laguna by the renowned cosmologist Professor Steven Weinberg with the title 'ORIGINS'. This Nobel laureate and celebrated author of *The First Three Minutes* gave us quite a few problems. For a start, he insisted on staying at the same hotel as the heads of state, on the grounds that, being Jewish, he was at risk from terrorists and therefore needed the same level of security as the heads of state. Quite naturally, it occurred to the chief of security that, if Weinberg really were such a magnet for terrorist attempts, he was himself a security risk and would need to be accommodated as far away as possible from the heads of state.

When he arrived at the University of La Laguna and saw German flags at the entrance and he said he would not speak in the presence of 'Nazi' symbols. This was a serious upset because the auditorium was full to the brim. We told him firmly and politely that there were certainly no Nazi symbols present, and that Germany was a signatory country to the international agreements that supported the Observatories about to be inaugurated. Regardless of these hiccups, Weinberg went on to deliver a brilliant talk.

The last straw came when, now back in Texas, Weinberg received the previously agreed emolument for his talk; he rang us and caused a scene. In the background, we could hear him arguing with his wife, who was shouting that the money received was not enough and was an insult to such a prominent personality, and that we needed to remember that we were dealing with no less a person than a Nobel laureate. Eventually he calmed down and accepted the payment.

On the island of Tenerife, the King and Queen of Spain, accompanied by the other dignitaries, inaugurated the Instituto de Astrofísica de Canarias in La Laguna and re-inaugurated Teide Observatory. There was a delightful moment at the ceremony held at IAC headquarters in which the then Prince of Asturias, the adolescent Don Felipe de Bourbon, was named 'Honorary Astrophysicist of the IAC'. We knew of his childhood love of astrophysics and his wish to have graduated in that discipline. It was a very solemn ceremony. A decorative dais had been erected for the heads of state and royal families to be seated. Facing the dais, in the patio of the Library, there was a stand for the invited guests. The Prince was handed a certificate of the title and a decorative symbolic sash of the kind worn by medieval astronomers, designed with the help of Professor Juan Vernet, a distinguished Spanish historian of mediaeval astronomy.

Following the Prince's words of gratitude, decorations of the Order of Civil Merit were bestowed on the most significant persons associated with the

Fig. 13.1 The author going over the notes for his inauguration address under the bright sun of Teide Observatory. (Credit: IAC)

origins of the IAC. I received the title of Commander in Chief, and Rafael Arnay, José Antonio Bonet, Maximino Galán, Manuel Vázquez, and Carlos Sánchez were made Commanders. Much to the chagrin of his father, Carlos Sánchez refused to step up to receive his award. All this business with 'bits of tin', as he called it, seemed to him to be a lot of archaic nonsense.

The facilities of the new Institute of Astrophysics were toured and a symbolic plaque unveiled, which in our case was a bronze cube with allusive texts in Spanish and English, similar to those located at each of the Observatories. They remain there today to commemorate these events.

The visitors were then shown an exhibition, 'Astrophysics in the Canaries', which displayed the present and, above all, the promising future of this science in Spain, due in large measure to the islands' Observatories. In the Library they signed the Roll of Honour, which was also inaugurated. As he was leaving, Don Juan Carlos, who was the last one to sign, turned and said to me sardonically, 'And the pen, Paco? Careful with that lot! If you knew them like I do ...'. The object of his humour was a golden baroque quill, purchased for the occasion. I reassured him that it was safe in my jacket pocket, as an historical memento of the IAC. Another example of the bonhomie of the King occurred during the exhibition that we had prepared on Astronomy in the Canaries when, seeing how I was struggling with my

Fig. 13.2 The author receiving the medal of Commander in Chief of the Order of Civil Merit from King Juan Carlos I during the inauguration of the IAC in La Laguna. (Credit: IAC)

English, he offered to act as interpreter so that his companions could follow my explanations.

The tour ended in the workshops, where, because it was such a large area (half empty at the time), all the guests could be gathered together. Today it is crammed with machinery and its capacity has had to be expanded threefold. There, the Nobel laureates, authorities, distinguished scientists, and others met all the staff of the IAC, which then numbered less than a hundred. To conclude the visit, the monarchs and heads of state posed for a 'family photo' with just the IAC staff.

After the inauguration of the Central Headquarters of the IAC in La Laguna, there was a gala feast in Puerto de la Cruz, offered by the King and Queen of Spain to the personalities who had attended, and to which every member of the IAC was invited. It was a glamorous event, which Canarian 'high society' scrambled to attend. It was amusing to see habitually untidy astrophysicists donning smoking jackets and workshop staff not in their customary overalls. But most amusing was the sight of my secretary, Campbell Warden, at the entrance to the emblematic Hotel Taoro, making an obeisance to each sovereign that arrived. Carried away by the occasion, he later

Fig. 13.3 'Family photo' of IAC staff with the heads of state attending the inaugurations of the IAC and its Observatories at the door of the Mechanics Workshop of IAC headquarters in La Laguna. (Credit: IAC.)

commandeered the official chauffeur-driven car that had been assigned to me as IAC Director for the event, so I was left to thumb a lift home with my wife to dress for dinner so as not arrive late at the feast. Many have been given their marching orders for far lesser offences.

That same night a ship carrying IAC people, journalists, and invited guests set sail for La Palma. From what I am told, the celebrations continued in full swing.

Roque de los Muchachos Observatory greeted the day of the inauguration with its best smile. The sun was radiant in a perfectly clear sky; there was no wind, and a green-yellow mantle of *codeso* flowers impregnated the air with fragrance, while in the distance, way below, a dark blue sea served as an aquatic pedestal for the island. The white domes of the telescopes contrasted with a sky saturated with blue. To add to the scene, enormous 'flags of the cosmos', designed by César Manrique, undulated in a light breeze, to the accompaniment of music composed for the occasion by Carmen Hernández.

The main ceremony took place in a rustic-style amphitheatre, built on the mountainside expressly for the occasion. Speeches were made, in order, by the President of the Governing Council of the IAC, the President of the Canarian Government, the Director of the IAC, the President of the International

Fig. 13.4   Close-up of the 'family photo'. (Credit: IAC)

Scientific Committee, and His Majesty the King, who declared Roque de los Muchachos Observatory duly inaugurated.

The visitors, beginning with the royalty, heads of state, and principal authorities, took the opportunity to tour all the telescopes. At the telescope of the Swedish Academy of Sciences, the King of Sweden conferred on me the medal of Commander of the Order of the Pole Star.

The ceremonies were a great success. The day ended with an open-air meal, at which local Canarian fare was served, and the participants mingled and talked among themselves. The VIPs retired to rest while the other participants roamed about the Observatory.

For operational purposes, the royalty and heads of state were transported by minibus within the Observatory. It fell to me to accompany them. They were happy, excited, and talkative. They moved from one window to another like children on an excursion. I find that nobody has forgotten those days, for over the years I have chanced to meet quite a number of the participants— some heads of state, others scientists—and without exception they tell me that those days they spent in the Canaries were unforgettable.

The final event took place in Santa Cruz de La Palma, in the Convent of San Francisco, where a recital was given by the soprano Victoria de los Ángeles.

Moments before the recital, each of the monarchs and heads of state who had attended the inauguration of Roque de los Muchachos Observatory planted orange trees as a memento. That orchard still stands as an added tourist attraction of La Palma.

The Convent of San Francisco was founded in the sixteenth century, the first of its buildings dating from 1508. At present, after its restoration, the Royal Convent of the Immaculate Conception houses the Island Museum of Fine Arts, Natural Sciences, and Ethnography; the General Archive of La Palma, and the Island Library José Pérez Vidal. The edifice stands out stylistically for the sobriety of its lines and for the Franciscan austerity of its ornamentation. It is the only convent on the island of La Palma that has been conserved in its entirety.

At the time, Javier Solana, Professor of Physics at the Complutense University of Madrid (which is how I had previously come to know him personally), was the Minister of Culture. I went to see him to seek his help in adding cultural content to the more important ceremonies that we were planning. He was most receptive and took a hand, as we had hoped he would, in ensuring that some cultural residue of the events would remain on the island. It was especially thanks to him that his ministry became involved in three matters: the restoration of two emblematic buildings in Santa Cruz de La

Fig. 13.5 General view of the inauguration ceremony of Roque de los Muchachos Observatory seen from the purpose-built amphitheatre, with the presidential dais in the background. (Credit: IAC)

Palma (the Convent of San Francisco and the Salazar Palace) and assuming the lion's share of the costs of the exhibition 'Astronomical Instruments in Medieval Spain. Its Influence on Europe'.

This important exhibition, which was organized by Professor Juan Vernet, brought together a complete collection of Hispano–Arabic astronomical instruments from museums and institutions from all over the world. A lavish bilingual catalogue was produced that is now the envy of collectors. The exhibition showed the decisive influence of medieval Spanish knowledge on western culture and how it affected the renaissance discoveries that revolutionized our view of the Universe and opened the door to modern science.

The mansion of the Salazars, which housed the exhibition, was built by Ventura de Frías Salazar, a knight of the Order of Calatrava and alderman of the Palmeran Council from 1631 to 1642. It was constructed in a baroque style consisting of ashlars and ornate carved figures in its façade, the rest of the building conforming to traditional Canarian architecture. Of particular note is the imposing Mudejar-style roof. The palace is now a beautiful cultural centre in which exhibitions, conferences, meetings, and other events are celebrated all year round.

Of all the ministers and politicians that participated in the 1985 inaugurations, the one who showed most interest in the science under way and being planned at the IAC observatories was the Minister of the Treasury Department, Miguel Boyer. He was curious about the telescopes and other instruments and asked endless questions. He was clearly most content. He was the only minister to seek me out to congratulate me. I remember well how, after the final ceremony, in the Convent of San Francisco, he came to bid me farewell and to tell me that, of all the research centres he had visited in Spain and abroad, the IAC and its Observatories were the most impressive and augured well for the future of the country. He added that it would be a great service to the country to fight to guarantee its growth. He insisted that, should any economic problem arise, I contact him directly. Hearing the Minister for the Treasury Department say such a thing was like a direct communication from God Himself, a dream come true for any recipient of public funds. Three days later he was removed from office! It was too good to last. Boyer was an aficionado of physics, and especially cosmology. When I visited him years later at his office as President of Banco Exterior de España, I saw his bookshelves lined with books, a few of them on economics, but the majority on physics and astrophysics.

I was interviewed many times by newspapers, radio, and television, but one question by a journalist stands out in my mind. The journalist, impressed by what he was seeing and seeking to flatter me, asked, 'Are you pleased with this wonderful culmination of the IAC?' I remember his surprise when I told him, without pausing to think, that what he was seeing now was merely a decorative doorway to many things to come. The good times, I continued, were just beginning and he would later witness the many great things *el Astrofísico* would accomplish in the future. It was no idle boast, and the facts speak for themselves. Perhaps the IAC's most distinguished achievement so far has been the design, construction, and commissioning of the Gran Telescopio Canarias (GTC), more of which later.

As the inaugurations drew to a close, I was completely exhausted. They were packed, strenuous days with too many responsibilities heaped on my shoulders, what with the complicated inaugural ceremonies, with the added stress of constant toing and froing between Tenerife and La Palma. It is easy to imagine the fatigue that all this implied for those working both in the wings and centre stage. There had not been a single moment to take a breather and forget all the hassle, so my wife and I took the opportunity to avail ourselves of seats on one of the fleet of aeroplanes laid on by the government to transport the many persons involved back to Madrid and join the exodus for a well-earned holiday.

Looking back on those crowded and magnificent events and knowing the fruit they bore, one is satisfied that the immense efforts of all involved have been well rewarded. If we reckon the coverage generated by the astrophysical inaugurations of 1985 by the yardsticks used by publicity agencies, we would be astonished at the resulting 'astronomical' sum. But the really important point is that the originally quixotic dream of making astrophysics flourish in Spain to find its place in international Big Science was now becoming a reality.

Since then, the Canarian Observatories have not stopped growing and have established themselves as privileged sites for the most advanced telescopes to become a worldwide point of reference. Since their inaugurations the Observatories have continued to receive visits from personalities from the worlds of science, culture, royalty, politics, finance, etc., and people from all walks of life. Most importantly, however, we have always particularly encouraged visits by children and youth, who are builders of the future.

14

The Sky Law (1978–2017)

At the behest of the Canarian Parliament, in 1988 the Spanish Parliament passed the Law for the Protection of the Astronomical Quality of the Observatories of the IAC, more commonly known as the 'Sky Law'. The law came into effect by royal decree in 1992 with the approval of regulations governing its application. These regulations, which we had been years campaigning for, converted the Observatories into an 'astronomical reserve' that was unique in the world, and legally protected against contamination from light, radioelectric emissions, and airborne contaminants. Aircraft were also forbidden from flying above the Observatories. The regulations were a key feature in maintaining the Observatories as a magnet for the most advanced astronomical instruments.

Before the passing of this law we had followed a hazardous path. Even before the signing of the multinational Treaty of Cooperation in Astrophysics, we saw the need to protect the skies above the sites where so many advanced telescopes were headed. The summit of the island of La Palma had been selected for the largest telescopes precisely because of the absence of light pollution, unlike the now heavily light-polluted Tenerife. However, with economic prosperity came an increase in electrical power consumption on La Palma, and public exterior illumination was beginning to get out of hand.

In 1978 we tried to solve the problem via 'municipal ordinances' through the good offices of the Island Cabildo. The Cabildo's President at the time, Mr. Carrillo Kábanas, saw with clear judgement and intelligence the benefits that would accrue to the island once the Observatory was built at Roque de los Muchachos.

© The Author(s), under exclusive license to Springer Nature Switzerland AG 2021
F. Sánchez, *The Rise of Astrophysics in Modern Spain*, Astronomers' Universe,
https://doi.org/10.1007/978-3-030-66426-8_14

Fig. 14.1 Satellite view of Europe and the Middle East in Google Maps. Light contamination from major cities is clearly seen. (Credit: NASA)

It was impossible to reach an agreement among all the municipalities on the island. I made a peregrination through all the towns in order to try and convince municipal authorities of the advantages that astronomy would bring for the island. To achieve this, I insisted, it was vital to maintain the clarity and darkness of the sky, a task that was wholly compatible with other local aspirations through intelligent lighting. But there was no way. In truth, I could understand their point of view. No so long before there had been neither illumination nor public electricity. They knew how large and prosperous cities throughout the world glowed; and now, to ascend a gorge and be able to contemplate from afar how their own towns similarly glowed was for them the greatest sign of modernity and progress.

We concluded that the only solution was a parliamentary law in which protective measures were accompanied by large subsidies for public lighting in accordance with the measures laid down. In the early eighties we began a new battle to persuade the politicians of all parties to support the law. This very balanced law was backed by the International Scientific Committee of the Canarian Observatories, the International Astronomical Union, and the International Commission on Illumination.

I refrain from wearying the reader with a detailed account of the stony path trodden. What is important is that we finally reached our goal in 1988, and the Sky Law is still a point of reference worldwide for the protection of the night sky. It has served as a model in Spain and other countries for similar protective measures. It is also the basis for night-sky protective initiatives and recognition of their cultural and touristic value by the Starlight Foundation (Chap. 23).

Fig. 14.2 An example of how light contamination drowns out starlight. (Credit: IAC)

Briefly, the law covers four fundamental aspects:

- Light pollution: the law regulates exterior illumination on the island of La Palma and that part of the island of Tenerife directly facing La Palma.
- Radioelectric contamination: levels of electromagnetic radiation are established to prevent interference with the scientific equipment of the Observatories that might harm observations.
- Atmospheric contamination: activities that might degrade the quality of the atmosphere at the Observatories are controlled.
- Air routes: air traffic is regulated to prevent flights over the Observatories.

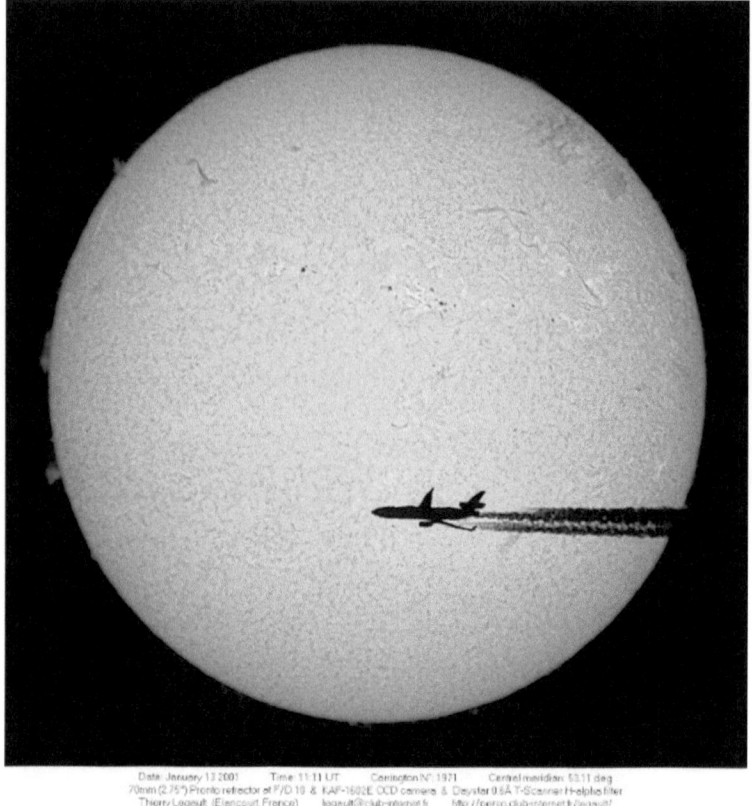

Date: January 13 2001   Time: 11:11 UT   Carrington N° 1971   Central meridian: 53.11 deg
70mm (2.75") Pronto refractor et F/D 10 & KAF-1602E CCD camera & Daystar 0.6Å T-Scanner H-alpha filter
Thierry Legault (Elancourt France)   legault@club-internet.fr   http://perso.club-internet.fr/legault/

Fig. 14.3 Air routes can also affect observations of the Universe. Under the Sky Law, Air traffic is prohibited over the Canarian Observatories. (Credit: Thierry Legault/ Elancourt, France)

The most noteworthy benefits of the law include protection of the night sky for both astronomical and cultural purposes, protection of the environment and biodiversity, and energy saving.

We realized right away that, if this law was to be observed, permanent technical vigilance was necessary, in combination with a continuous programme of social awareness and education. For this purpose, in 1992 the IAC created the Technical Office for the Protection of Sky Quality (OTPC), which serves to orientate citizens on how to conform to the law and oversee its application. Special talks and courses were given to public illumination technicians of the Canarian municipalities. Much emphasis has always been placed on outreach in the form of talks, publications, videos, and social networks. Every attempt has been made to collaborate with firms that produce and install lighting fixtures, and we have succeeded in persuading some of these firms expressly to redesign new non-contaminating screen models.

Little by little we have managed to persuade the authorities and the general public of La Palma of the benefits of this innovative law, which has shown a measurable result in terms of economic markers. And attitudes have changed from viewing night-sky protection as a burden to considering it a privilege that adds a distinctive note to the island. At the commemoration of the thirtieth anniversary of the Sky Law, the Mayoress of Llanos de Aridane said, 'We must be the communication channel broadcasting the necessity of protecting this treasure of our land. We must make good use of, exploit, protect, and cherish it.'

The main obstacle has always been keeping external lighting within the regulations. When funding an illumination project, the town halls, whether by custom or through vested interests, do not always monitor the correct application of the regulations, especially where new lighting projects are concerned. This, combined with pressure from vendors, has meant that the OTPC has always had to issue official reports. But, because the Sky Law does not incorporate any coercive measures, it is the local authorities in charge of electrical and environmental matters that must impose any fines. And since the fines always arrive late—or never—the deterioration of sky quality is beginning to become a serious issue in certain zones.

Similarly, the introduction of LED lighting, which is harmful to the health of both humans and local fauna, is undermining the application of the Sky Law. This is another front that the OTPC is having to cover with its meagre resources. For its part, the IAC has been promoting the adjustment of the current legislation to address the problem of these new lamps in order not only to avoid the serious impact of blue LEDs on the environment but also to regulate the use of LED sources in the ambit of the application of the Sky Law. The procedure was initiated in 2015, but Royal Decree 580/2017, enabling the new regulations, was not published in the Official State Bulletin until 12 June 2017. It is now to be expected that the Canarian Government will be rigorous and swift in its application.

Unfortunately, we find ourselves immersed in the intricate world of laws and regulations, in which regulations of the European Union, the state, the corresponding autonomous community, the municipalities, etc., intertwine and overlap. A case now occupying us is just one example more of the tangle that binds and asphyxiates us. To our Sky Law (Law 31/1988 of 31 October and Royal Decree 243/1992 of 13 March which regulates it) must be added a number of other pieces of legislation. These include: Law 37/2007 concerning air quality and protection of the atmosphere, and Royal Decree 1890/2008

Fig. 14.4 Nocturnal panoramic view of the protected Roque de los Muchachos Observatory under a star-studded sky and an overarching Milky Way. (Credit: Daniel López: IAC)

which regulates it; regulations concerning energy efficiency in external lighting installations and their complementary technical instructions; guides to the regulation of energy efficiency in external lighting installations; 'reference criteria concerning lamps relative to %FHS (according to regulation 245/2009 of the European Commission)'; rules governing sports and leisure facilities of the Upper Council of Sports; basic rules for radioelectric facility projects; etc. That is how things are done in Spain.

In spite of all this, we at the IAC are optimistic, and our OTPC colleagues tell me that the time will come when the quality of the skies above La Palma will be an astronomical resource valued not only by scientists but also by the islanders. Possession of this jewel will be a source of pride to them, and it will be they who strive to keep it. Let us hope the OTPC is right.

15

The IAC: A Dream Come True (1971–Present)

So far, I have recounted a wide variety of things that happened in the sixties, seventies, and eighties of the last century, and I think the moment has come to hang the fruits harvested in these decades on the tree that supports everything: the Instituto de Astrofísica de Canarias. Bear in mind that those years saw the successful and peaceful transition from dictatorship to democracy in Spain, and it is no coincidence that our transition from classical astronomy to modern astrophysics should have occurred over the same period. I also think that the following chapters will make more sense if we dedicate this one to the IAC itself in order to summarize the concepts, and the whys and wherefores of this organization.

It began as a yearning, a dream. As soon as we began the site testing and compared our data with those from the best sites where others were working, we appreciated the undoubted excellence of the mountaintops of the Canaries for modern astronomy, and my wife and I decided to stay here and dedicate ourselves to ensuring that this 'natural resource' should be put to good use for the benefit of science and our country.

But how to convert these germinal stirrings into a reality? Of course, to attain our goal, to reach our destination, much needed to be done by people with drive, vision, means, and an adequate organization.

What could be done from Izaña, perched on the top of Tenerife, so far from Madrid and the rest of the world? I tried first of all to convince the Director of Teide Observatory and enthuse him with our vision. I kept on at him repeatedly, as my many letters and reports to him confirm (I still have these in my possession). No luck there, so I rolled up my sleeves and set about the task myself, with all the optimism of a young tyro and the passion of a believer.

© The Author(s), under exclusive license to Springer Nature Switzerland AG 2021   **137**
F. Sánchez, *The Rise of Astrophysics in Modern Spain*, Astronomers' Universe,
https://doi.org/10.1007/978-3-030-66426-8_15

Many were the wiles we had to employ here in Spain to secure the birth, setting up, formalization, and growth of the IAC and astrophysics in the country. We had to know how to trim our sails with astuteness when the winds were propitious, and slacken them to avoid adverse winds, all the while keeping a firm, determined grip on the tiller, never deviating from our set course.

For a start we had to commit ourselves to achieving four preliminary objectives:

- Acquire a voice, independence, and control,
- Attract modern telescopes then being constructed to the Canaries,
- Raise the banner of astrophysics, and
- Begin without further delay to train young Spanish astrophysicists.

We achieved all that!

Those objectives were attained the seventies. During that time the basic nucleus of personnel was formed that would carry the great project forward.

To fly high one must first take off. We had to have a voice and control. In the Spanish academia of those days, that meant that obtaining a professorship was an essential prerequisite. So, with that aim, the first thing to do was to campaign for the creation a chair in astrophysics, quite an exotic demand at the time. But before I finally got the chair, the Director of Teide Observatory threatened to expel me from the Observatory if I dared present myself as a candidate for the professorship on the grounds that my activities at the Observatory would be incompatible with the holding of such a chair! And this from no less than the person who at the time was Professor of Astronomy and Geodesy and Dean of the Science Faculty of the Complutense University, Astronomer at the National Observatory of Madrid, Director of the School of Topography, and permanent secretary of the Royal Academy of Sciences—on top of being Director of Teide Observatory.

At that time *oposiciones* for professorships were all held in Madrid. During my examination in the auditorium there was a handful of young physics lecturers who had been rejected by the tribunal, but who stayed behind curious to know who this unknown Doctor Sánchez from the Canaries was and how he would comport himself. At the end of the many long examinations typical of the period, I succeeded in demonstrating my knowledge and capabilities to two of the members of the tribunal and, finally, to the professor of Astronomy of Barcelona, Professor Orús, who tipped the balance in my favour. I got my professorship.

The seven or eight persons who then constituted astrophysics in the Canaries presented me with a gown, cape, and biretta for my professorial attire, which I displayed for them standing on a table in the laboratory during the celebration they had planned. It was a significant and charming memory that I hold dear.

Once I became a professor of astrophysics, things began to change; we could now reach forums where our voice could be heard and listened to. At last, we could set about doing what we needed to do.

In 1971 I was elected a member of the International Astronomical Union after being nominated by Commission 21 of that organization, rather than through the National Commission of Astronomy, as would have been the normal procedure. The National Commission of Astronomy was therefore forced, by virtue of its own statutes, to accept me as a member, and the Presidency of the Government had to give me the right to vote in that consultative body of the government. We were beginning to gain a presence and a voice.

At the beginning of 1972 I was summoned by the Director General of Universities and Research to the ministry to 'study matters relating to Astronomy and Astrophysics', together with three professors of astronomy (Torroja, Orús, and Cid) and Padre Romañá, President of the Alfonso the Wise Council of CSIC. We were building up our presence and voice.

Also in 1972, the European Space Research Organisation (ESRO, the forerunner of ESA) nominated me a member of its Solar System Working Group, which supervised all European space mission projects.

The time had come to tackle the question of creating an organization that would sustain all the astrophysical activity about to become a reality in the Canaries. It would need to be a structure with the capacity to receive the staffing and material means necessary for what we had in mind, and it would need to be managed efficiently. But it first of all needed to be invented and built up step by step.

In 1973 the University of La Laguna created its 'University Institute of Astrophysics', to which Teide Observatory would eventually be answerable, and I was appointed as its Director by ministerial decree. Professor Torroja abandoned his post, but without formally renouncing it.

At that time, the matter was raised of creating a 'National Institute of Astronomy', a large national organization, controlled by CSIC, that would centralize and manage all astronomical resources from Madrid so that the same people could continue to meddle in everything. This manoeuvre was halted in its tracks in 1973 when the Inter-island Association of Cabildos of the Province of Santa Cruz created the 'Institute of Astrophysics of the

Association of Cabildos', to which were ceded the lands for building Roque de los Muchachos Observatory.

We could now set things in motion. We could now act to put the Canarian skies to good use (our ultimate goal). The first step was undeniably the articulation of a genuinely international collaboration, rather than the mere cession of lands on which foreign organizations could set up their telescopes. We were not going to fall victim to the colonialism typical of the American and European observatories in Chile, or the German observatory in Calar Alto in Spain, which were colonial-style foreign enclaves. It was not enough merely for latest-generation telescopes to arrive in the Canaries; the science done with these telescopes had to be properly shared with Spanish scientists. The owner institutions would also have to help us set up a Spanish research centre.

For our part, we had quickly to train home-grown scientists and technologists so that they could take their place at the telescopes from day one. The crux of the matter was to avoid the type of scientific colonialism just mentioned and to be able to derive palpable benefits for the environment. I described earlier how we managed to persuade those who in due course would

Fig. 15.1 Land near the old road to La Esperanza when it was allocated by the Tenerife Cabildo for building the University Institute of Astrophysics of the University of La Laguna, precursor to the present IAC. (Credit: F. Sánchez)

Fig. 15.2 The former University Institute of Astrophysics. The site is now occupied by the Faculties of Physics and Mathematics in Avenida Astrofísico Francisco Sánchez. (Credit: IAC)

become our colleagues to sign the Treaty concerning Cooperation in Astrophysics, which is still in force today.

With the decks now cleared, more ability to act, and knowing exactly where we were headed, the initial team (no more than fifteen persons, most of them still without permanent posts) got down to work. We were young, and the challenges, although difficult, were beautiful, filled with hope, and excitement. We began more and more to believe in the possibilities.

As we were so few we worked through general meetings until we became aware of the time being wasted in this way, so we left management to the fewest possible members of the team so that the rest could do science and technology, which was more important. Thus began what was later to become the IAC's Management Committee.

Another important step was taken in 1975 when we created a consortium to bring together those entities that had something to contribute to the project; we also incorporated CSIC into the new consortium in order to emphasize its national character. We invited the following persons to take part in the drafting of the document: Fernández Caldas, then Chancellor of the University of La Laguna; Rafael Clavijo, then President of the Inter-island Association of Cabildos; Primo Yúfera, President of CSIC; and Mayor Zaragoza, then

Minister for Education and Science. The Instituto de Astrofísica de Canarias was born by drawing together all these entities through a 'cooperation agreement', signed in the presence of the Minister for Education and Science, by the University of La Laguna, CSIC, and the Provincial Inter-island Association of Cabildos of the Province of Santa Cruz de Tenerife. With this agreement, all astrophysical activity in the Canaries was integrated under the command of a single entity. This consortium model would serve as a conceptual basis for the Astrophysics Law of 1982 mentioned earlier (see Chap. 12).

With this agreement we advanced another step in our strategy of consolidating our nascent Institute and setting it on a formal basis. Our aim was to become an entity capable of managing the Canarian Observatories and all the activity on the islands in this branch of science. We were ever wary of the Spanish weakness for territorialism, which would result in astrophysics being split up piecemeal with constant squabbling among institutions—always an endemic tendency in Spain.

In the summer of 1975, a select group of outstanding Spanish physicists held a week-long meeting in Santander under the auspices of the Menéndez y Pelayo University to analyse the state of physics in the country and to try to coordinate plans for its future. I remember Javier Solana, who was later to rise to dizzy heights in politics, being among them. As there were at the time a great number of theoretical physicists, there was a noticeable tendency in the discussions to slant the development of astrophysics towards theory. I had to work hard to convince them that, with the arrival of powerful and advanced telescopes, it was necessary to develop observational astrophysics and its attendant technologies. I argued that the theoretical side would develop without hindrance, given the deep-rootedness of theoretical physics at our universities, which would ensure its economic development. I can remember the support I received from Nicolás Cabrera (the son of Don Blas), who had just returned to Spain and was at the time setting in motion the teaching of physics at the newly created Autonomous University of Madrid.

Another matter deserving mention is the formalization of the concept of the Canarian skies as an exploitable natural resource at a gathering of economists. The First Economics Studies Workshop—organized by the Canarian Council (the pre-autonomy government of the Canary Islands), Banco de Bilbao, and the University of La Laguna—was celebrated during 9–11 April 1980. The title of my talk, 'The skies of the Canary Islands as a natural resource of recognized scientific interest', expressed with clarity the concept that was to serve as a permanent reference. I gave a summary of what had been achieved to date and a future projection of the IAC. I began:

It comes as a surprise to say that the skies, even those of the Canary Islands, may be seen as a natural resource in the commonly accepted meaning of that term in economics. However, the circumstances we found in exploiting astronomically a certain zone of the summits of La Palma and Tenerife have led us to understand that these correspond to the characteristics of those bounties offered by nature that are desirable to man at a particular determined epoch.

I rounded off with:

The Agreements achieved will be both fruitful and advantageous, not only for our science and reputation but also in ensuring the continuous influx of European lecturers and researchers, a guarantee of the permanent injection of western culture in the Canaries that is so vital for everything, especially in these times. Moreover, Spanish researchers will be able to avail themselves of modern working tools: highly qualified positions will be created, and the road system and power grid of La Palma completed. Without a doubt, all this will contribute to the scientific, technical, and even touristic development of the region.

At the time I was Chancellor of the University and my understandable preoccupation with the imported independence movement that was so active at the time in the Canaries was evident.

We have had to wait no less than 38 years to confirm my assertions with economic data. At the beginning of 2018 a well-written and documented study bearing the title 'The Economic and Social Impact of Astrophysics in the Canary Islands' was published by Professor Juan José Díaz of the Department of Economics, Accountancy, and Finance at the University of La Laguna.

One must always make the best of circumstances. Earlier, we saw the serious problems that arose with the Ministry of Defence in our struggle to prevent the installation of military radars from interfering with the building of Roque de los Muchachos Observatory. As a consequence of that confrontation, we ended by establishing cordial relations with the civil servants of the Presidency of the Government, who were thus able to see for themselves and help us in what we were seeking, and understand how it would be beneficial for the country. That was how, when the doors of the Ministry for Universities and Research were closed to us (as described earlier), we were taken under the wing of the Ministry of the Presidency. This was a great piece of luck, for we had gained a much more protective umbrella and source of funding. The Treaty of Cooperation in Astrophysics was signed in the name of Spain by Mr. Pérez Llorca, Minister of the Presidency of the Government, and the funding to cover the agreement came from that ministry.

The definitive step in obtaining the longed-for juridico–administrative consolidation of the IAC was taken with the re-creation of the Instituto de Astrofísica de Canarias by a royal decree–law in 1982. The newborn IAC was answerable to the Ministry of the Presidency of the Government, the Minister of which chaired the IAC's Governing Council. This law endowed the IAC with the capacities and self-management necessary for us to carry out our tasks of administering the internationalized Observatories of the Canary Islands, doing research, developing and transferring technology, training researchers and technologists, doing science outreach, and promoting the environment. To fulfil these functions, we thought out and designed a simple and efficient organization to be set in motion immediately. All of this constitutes the Instituto de Astrofísica de Canarias of today. With this agreement we advanced another step in our strategy of consolidating our nascent Institute and setting it on a formal basis. Our aim was to become an entity capable of managing the Canarian Observatories and all the activity on the islands in this branch of science.

The eighties covered what we might call the third stage of this great project, which was the juridico–administrative consolidation phase of the IAC in 1982 and its introduction to society through the widely publicized inaugurations of 1985. This stage marked the IAC's scientific, technological, educational, and cultural consolidation. The initial goals culminated with this stage. It was then that I saw that I could, and should, focus all my efforts on directing the IAC; I left my teaching at the University and, even though I could have combined my chair with my directorial obligations to the IAC, I resigned my professorship. I also gave up research and disbanded my research team so that personal obligations should not get in the way of my duties as Director.

To recapitulate, the final years of the seventies and the early years of the eighties were a period of frenetic activity. We were fighting to ensure the arrival of telescopes without colonialist strings attached. But we were also training astrophysicists and initiating the teaching of astrophysics at Spanish universities. As if that were not enough, the chancellorship of the University of La Laguna fell to me at a particularly turbulent moment. We immediately tackled the up-hill task of legally formalizing the IAC. As a grand finale, we organized the inaugurations of 1985. At the same time, we had to obtain resources for the day-to-day and future running of the IAC. It was a period of great toil and many complications but an exciting one.

I shall now describe how the IAC found a physical home. In the middle of the last century, when a graduate from a poor family began to work and got married, he had to find accommodation in a small, often dingy, apartment. As he slowly climbed the career ladder and his status improved, the family would then move to better accommodation. Finally, if all had gone well, he would end up owning his own house in a good neighbourhood. It is curious to observe how the same thing has happened with the IAC. We shall see how the evolution of the successive headquarters of the IAC corresponded to the consolidation of its fortunes.

At the end of the sixties, as an already permanent adjunct lecturer, I asked for an office in the University of La Laguna and was only offered a kind of attic that had previously been used as a lumber room. I cleaned, painted, and converted it into a pleasant and comfortable workspace. This was the operational headquarters from which I started.

When I became a professor, the Instituto Universitario de Astrofísica was created and I was given a bigger space in the recently constructed Chemistry Tower. The headquarters were moved to this new site, with space for some offices and laboratories. The Institute's primitive workshop was set up in the basement, which is where our dedication to self-made instrumentation began to take shape. This space was to be the location where the present workshops of the University of La Laguna originated.

In 1975 the IAC Consortium was established, and the Island Cabildo of Tenerife donated a site in the former Carretera de La Esperanza, the road leading to the Teide National Park and Teide Observatory. The Cabildo itself funded a number of red-roofed uralite huts, of the sort that were being used for provisional schools. We set up the brand-new Institute right there in the huts: offices for astronomers, directorate and administration, a lecture room, and a computer centre (something completely new in the Canaries). Other huts were set aside for the mechanics and electronics workshops. We surrounded these simple, ugly, and cold buildings with pretty gardens and we furnished them simply and practically. When it was all finished it seemed like paradise to us. It gave us a sense of identity and cohesion. We were still very few at the time, perhaps twenty-five, and contact was direct and friendly. The work and scientific and technical discussions often continued into the night in the bars of La Laguna frequented by University people. I think of *El Buho* ('The Owl') as a nocturnal IAC. There you could always meet any of our astrophysicists when they were not observing up the mountain. We all look back on those days with nostalgia.

Our Instituto de Astrofísica, together with the Institute of Natural Products, which now bears the name of its founder, Antonio González, were the seeds from which sprang today's Science Campus of the University of La Laguna. The Campus now hosts the Faculties of Mathematics, Physics, Chemistry, Biology, Pharmacy, and the Higher School of Engineering and Technology and is located in the Avenida Astrofísico Francisco Sánchez (formerly the road to La Esperanza).

I mention an event that shows how *el Astrofísico* became implanted in Canarian society. When, at the beginning of 1980, the Instituto de Astrofísica was being moved to its new location in the south of La Laguna, the neighbours of Carretera de La Esperanza, which no longer led to either La Esperanza or to the Teide, having been truncated by the extension of the runway of Los Rodeos Airport, wrote a letter to the town hall earnestly requesting that their road be renamed Astrofísico Francisco Sánchez. The reasons they gave were quite charming: it would serve to 'remember this very kind gentleman who passed in front of our houses every day on his way to work'. After the usual lengthy procedures, in 1986 there was a curtain-drawing ceremony to reveal the new name in the presence of the authorities, the local police, a municipal band, etc., all very reminiscent of a Berlanga film. But the most moving part was that the neighbours had laid on food and refreshments in their homes, with an invitation to all those present to 'toast' the event in Canarian style.

Once the Treaty of Cooperation in Astrophysics had been signed and the IAC formally constituted with full legal rigour, our numbers began to swell and the huts became overcrowded. We needed more land and I went, once more, to see the President of the Cabildo of Tenerife, the intelligent, kindly, and visionary Rafael Clavijo, to request an extension to the plot we were occupying. To my surprise he refused. Instead, he offered us a much more extensive plot in a better location, next to the road to Santa Cruz via La Cuesta, so we went together to see it. On the way, he explained that his reasons were to do with urban development: he wanted to dignify the marginalized neighbourhoods in the southern part of La Laguna; if *el Astrofísico* were to be located there, he reasoned, things would begin to change. And that is actually what happened. Thanks to the vision of this politician, that zone is now home to the Museum of Science and the Cosmos, and the Faculties of Letters, Law, Economics, etc. There are pavements and public lighting, new houses, shops, and bars, and the Santa Cruz–La Laguna tram passes right through the neighbourhood. There is even an enormous statue of a half-naked, defiant Guanche in front of the IAC, similar in style to the Indian chiefs of western movies.

The present central headquarters of the IAC occupy an area of 22,000 square metres. Following the layout of the former huts, the departments are interconnected by corridors to combat the adverse climate of La Laguna. We obtained funding from the Presidency of the Government of Spain for the building works as part of the package of commitments acquired through the signing of the Treaty for Cooperation in Astrophysics. To ensure that the building work was properly supervised we appointed as permanent works manager Maximino Galán ('Chiqui'), mathematician and head of our workshops. Everything turned out well, proof of this being the minimal repairs and retouches that have been made over the thirty-year existence of the buildings.

At the beginning of 1985, half of the office space was empty, but it did not take long to fill it. Today, in spite of extensions made to the habitable space, the buildings are again fully occupied. Even the area above the canteen, originally designated as accommodation for visitors, has had to be converted into offices. That residence amply fulfilled its purpose of temporarily retaining at the IAC excellent astrophysicists from all over the world in order to provide contact with our young doctoral students. There are now around four hundred persons working in these crammed offices, including permanent and temporary staff, doctoral students, and visitors. A success story run riot.

Turning now to funding, this is a key element in all development and the foundation on which families, states, and businesses are built, and which drives all human activity. It also needs to be borne in mind that the IAC was born precisely into a Spain in transition towards democracy, with all the attendant political effervescence, all minds being focussed on the change of regime and the politicians preoccupied with the establishment of their political parties. Looked at from their point of view, in those times it must have seemed an absurdity, a waste of resources, to devote any effort at all to funding astrophysical research, a 'luxury for rich countries'. Scientific research was not in the least a matter causing any leader to lose sleep. Unfortunately, neither is it today. We are not the United States, where other criteria prevail and even entrepreneurs believe in the need to promote research, even if only because it helps them fill their pockets.

Through the Law of Astrophysics, the IAC now figured in the general state budget, but we had to ensure, year by year, that the portion corresponding to the IAC grew in the public budgets of whichever ministry it was answerable to at the time. We also had to budget for new personnel, and later better their job security and working conditions. Hence, it was not enough just to raise the banner of fundamental pure research; it was also necessary to add other, 'more useful', things. We tried our hand at comprehensible realities. One such reality was the inflow of money resulting from foreign investment in large telescopes and the funds needed for their continued use. Another was the visibility and prestige that having in the Canaries, and therefore Spain, so many major institutions from the most advanced countries would represent. Then there was the advanced technology that could be generated and funnelled for the benefit of local firms and the national economy. And finally, the development of astronomical tourism would boost and add complementary incentives to our 'sun and sand' touristic boom.

The most complicated and difficult challenge at every turn has been having to deal with our political authorities. I have had to negotiate with twenty-eight ministers, all the presidents of the Canarian Government, and an infinity of deputies, mayors, and councillors. To put it bluntly, being able to liberate myself from this martyrdom has been a great blessing of my retirement.

In order to sell our 'message', we had to make frequent appearances in the local, national, and international news media, a form of advertising that was free of charge to the various administrations. The Director's Support Team (see Chap. 20) took charge of this. Thanks to all this, some politicians have come to believe in 'our products', and funding has been forthcoming for the activities of the IAC—never enough to fulfil the wishes of all, but nevertheless a cause of envy at the University and other research centres in the country, which consider us as in some way privileged. Such is the way of the world.

Each year we have had to move heaven and earth to acquire a half-decent budget. In other words, we had to badger the politicians and senior civil servants of the day in Madrid and the Canaries, pester them to the point of tedium, and drive home our case with documents and more documents. We also had simultaneously to look outside in order to supplement the meagre sums apportioned to us by Parliament. It has been necessary at every turn to acquire extra money to supplement the budget in the form of aid, projects of every type, grants, etc. It is useful to know that for many years these extra-budgetary sums amounted to 50 per cent of expenditure. In spite of everything, we have never had sufficient funding to do all that we wanted to do; instead, we have had to be content with doing what *could* be done. How we thought with envy of those rich American philanthropists who have bequeathed millions of dollars to research!

But a truly worrying concern today—even more, if it were possible, than the funding shortages—is the continuing expansion of bureaucratic red tape in Spain, which has created a mountain of difficulties for management, with all its attendant unnecessary and suffocating administrative obstacles and hurdles. Many stories could be told of instances beyond belief where great opportunities involving millions of euros have been thrown away. This is not the time or place to go into that. What is certainly true is that Spanish legislation and administration, with its overlapping layers and duplications arising from the division of jurisdictions among the state, autonomous communities, councils, and local governments, including the town halls, constitute a costly and oftentimes paralysingly inextricable tangle. Sadly, however many new regulations are introduced, supposedly to stop corruption, the easier it seems to become for the shameless to perpetrate fraud.

To round off this chapter I describe the present situation of the Instituto de Astrofísica de Canarias. Despite its much lower budget, its scientific production, with an average of one publication per day in international peer-reviewed journals, is on a par with that of any other leading centre of similar size. The IAC has now established its presence in the international community and participates in outstanding astrophysical projects and networks, as well as in the corresponding forums of research and development policy. In the Canaries the IAC, along with all astrophysical activity, is strongly supported by the citizens and politicians. In fact, astrophysics is one of the main priorities of the Intelligent Specialization Strategy (RIS3) for the period 2014–2020.

The IAC has signed over seventy-five collaboration agreements, not only with institutions having telescopes at the Canarian observatories but also with many international consortia. To name each and every signatory institution would be prolix and tedious, but I mention some of the better-known ones, such as NASA, the Max Planck Institutes, Caltech and the University of California, the University of Cambridge. The University of Manchester, the European Southern Observatory, the European Space Agency, the Association of Universities for Research in Astronomy (AURA), and the Cherenkov Telescope Array (CTA). The complete list can be seen on the IAC's website and in documents published by the IAC.

To summarize, in the short period of time described in this book Spanish astrophysics has risen from the humblest of beginnings to rank among the best in the world. It has been transformed from a mere dream, a vain hope, into a tangible and noteworthy reality.

I never tire of repeating that all of this was achieved by a group of ambitious, united, and hard-working people in pursuit of dreams, yes, but dreams converted into milestones that have been reached year after year. It takes good people to do good things. As there have been so many of them—astrophysicists, engineers, and workshop, observatory, and administrative personnel—I cannot cite them all by name. Nevertheless, they have been key to the success that is the IAC.

16

Towards
Excellence in Research (1971–Present)

I now describe how, in its brief existence, the IAC has steadily worked towards high-quality, competitive research. I lay more emphasis on how this was achieved rather than the numerous milestones reached. There is little point in tiresomely itemizing a long list of scientific achievements. They are there for all to see in the most cited journals, international ranking indexes, and the IAC's website. Summaries of the most important scientific results are also published in the IAC's annual reports, and the Director's résumés in successive Governing Councils, and in IAC publications.

We began to research and publish right from the beginning, in spite of our lack of suitable means. Our output improved when foreign telescopes began to arrive, first with the Bordeaux Telescope, then the solar telescopes of the Fraunhofer Institute, and later the infrared telescope of Imperial College.

As we began to grow and diversify, we needed to find our niche. The Director had to know where the 'business' of astrophysics in the rest of the world was headed, which were the most promising lines of research in studying the Universe. We therefore sought the counsel of wiser (some of them not so wise) authorities in matters astrophysical. At the Institute we strove to keep abreast of the many 'perspectives' on the future of astronomy that were being published by the most advanced countries. The list is a long one, so I cite only a few of them here. The first such report to fall into my hands was *Prospettive dell'Astronomia Ottica*, published in 1968 by the Italian National Optics Institute. I later acquired all the reports published every decade by American and European institutions. I also tried to attend the meetings and symposia on future perspectives. And in 2004 I contributed a paper with the title 'The IAC: Its role in leading the development of Spanish astrophysics' to the series

© The Author(s), under exclusive license to Springer Nature Switzerland AG 2021
F. Sánchez, *The Rise of Astrophysics in Modern Spain*, Astronomers' Universe,
https://doi.org/10.1007/978-3-030-66426-8_16

Organizations and Strategies in Astronomy, in which I wrote of the Gran Telescopio Canarias and Roque de los Muchachos Observatory as a home for the largest telescopes. At the 2008 ASTRONET meetings, I discussed strategic plans for European astronomy and the outstanding contribution of astrophysics in Spain. At last we were on the map and could participate on an equal footing in the formulation of future science policy.

Research attains excellence when it adds to our scientific knowledge, otherwise it cannot properly be called research. The question is, how to make a research institute a centre of excellence. It is a never-ending and far from straightforward task, demanding a constantly critical attitude, staying permanently on one's guard, continuous examination of what is being done, and implementing corrective actions so as not to lose tone or rhythm. And this state of tension must be maintained at all times. There are numerous examples of what I mean in the minutes of the IAC's Management Committee meetings. I have found notes in my handwriting with ideas for improving the quality of our research that were written during various sleepless nights over the years. My last nocturnal vigil was in 2011.

But to go back a little in time, it was in the eighties, once we were certain that the most advanced telescopes would find their way to our observatories and the IAC was legally established, that we could focus more on the challenge of the internal organization of our research efforts, always with the aim of achieving excellence with the support of our great observing capacity.

It is not enough to have a clear idea of where one wants to get to; strategies must be devised that take into account one's strengths and weaknesses. We were aware that our main weakness was ignorance; after all, we were new arrivals in astrophysics. But our main strength was the Canarian Observatories. We needed to keep this in mind always since we depended on them in order to do competitive research.

In the autumn of 1983, only a year after having achieved the Astrophysics Law, we invited a group of renowned foreign astrophysicists with experience in heading research teams. This working group was formed by twelve foreign and four IAC researchers. Those of us representing the IAC were given the mission of keeping the discussion on track to ensure that the counsels offered, given the IAC's circumstances, were realistic and practicable.

We specifically asked them for a 'pragmatic and practicable document that would enable the directorate of the IAC to set in motion a well-founded programme of research of genuine interest'. Before they arrived in the Canaries, we had sent them clear and detailed information on how we had got to this point, what the IAC was legally and structurally, and what was currently being done in research and other areas of activity. They spent the first days visiting the Observatories and the Institute, meeting and talking to people. During the last part of their visit, in Mesa del Mar, they were boarded in a small and very comfortable hotel far from touristic centres. It was there that they had their discussions, reached agreement, and drafted the final document.

What I wrote in the preface of the small book that we published that contained the document just referred to reflected our attitudes and patriotic aims: 'The handful of enthusiastic persons, which is the IAC, openly and humbly endeavours to continue to improve its research work and fulfil its social duty to deliver knowledge and technology directly to society, and takes this opportunity once again to seek help from the national scientific community'.

This document, published in Spanish and English with the title *Recommendations about the Future of Research at the IAC*, which we referred to among ourselves as the 'Book of Wisdom', did not fall on deaf ears. It is also an historical document, for it summarizes what the IAC was at the time and outlines what we wanted it to become. It was our bible for a long time. If a comparative study is made at some time of what the IAC has become, the influence of that little book will be evident. Even in such specific matters as the most appropriate staffing levels for an institution such as the IAC, with such a heavy observational burden, the book served as a point of reference. The advice it gave was: for each researcher, two technicians and 0.5 administrative staff. I like to think that, thanks to having always attempted to maintain these staffing proportions, the IAC has not yet descended into total bureaucratization, in spite of being ever more pressured by national and local administrative regulations.

The manual gave advice on the most fruitful lines of research for the IAC to follow, and even named the most suitable persons and research teams from all over the world with which to form collaborations.

With this External Assessment Committee, we created an unusual (for Spain) precedent. It took many years for this system, so effective and normal at other latitudes, to be adopted in varying degrees at other research centres in our country. But at the IAC calling on external teams of experts to make critical evaluations, suggest corrections, and offer advice has since that time become an institution in itself.

This external assessment committee became enshrined in the early Royal Decree concerning the Structure of the IAC as the Research Assessment Commission (CAI after its Spanish initials). The CAI, the decree states,

Will be constituted of persons of recognized scientific standing designated by the Governing Council and will be the senior consultative and assessorial body of the Institute for the purpose of orientating its policy with regard to scientific and technical research, and the programming of its activities in a coordinated manner in accordance with the prescriptions of the National Plan for Scientific Research and Technological Development.

To avoid cronyism and favouritism arising from too frequent well-received visits to the Canaries, new members would be regularly welcomed onto the Commission.

To give a concrete idea of what we are talking about and of the rigour with which we proceeded, I give as an example the constitution of the CAI that met in 1996. The Commission then comprised Margaret Burbidge, Steven Beckwith, Cornelius Zwaan, Gerry Gilmore, Martin Rees, and Guido Münch—all astrophysicists of great repute.

Also members were the secretary of the National Plan for R&D and a representative of the National Commission of Astronomy. The objective was, in this manner, to combine IAC research with Spanish science policy so as to unite the astrophysics community of the whole country.

The meetings were meticulously planned, and the CAI members were provided with full information before their arrival, and sufficient time was given to allow the Commission members to talk with whomever they wished at the Institute during their stay.

One of their recommendations was to allow researchers the freedom to work on the topic of their choice. In my opinion, a small centre such as the IAC with scant resources ought to concentrate on a small number of lines of research in order to avoid attempting too much and achieving too little. This recommendation, not to everybody's liking, has resulted in the Institute covering too wide a spectrum of topics: the Universe and cosmology, galaxies, stars, the sun, the interstellar medium, planetary systems, atmospheric optics, high spatial resolution, the design and construction of telescopes, and optical and infrared instrumentation for terrestrial and space astrophysics. I would have preferred the IAC to be a clear world leader in one or two of these topics.

The CAI report, accompanied by a 'Plan of Action' prepared by an ad hoc working group, would always be presented to the IAC's Governing Council in order to put the CAI's advice into action.

It pains me to report that it has not been at all easy to keep this useful external assessment committee intact. Incredibly, we have had to fight many battles to prevent it from being disbanded as the result of successive legal regulations that try to 'normalize' the IAC. These battles have taken place at all political levels with successive governments of Spain and the Autonomous Community of the Canaries. Finally, the CAI was given the mission of appointing the Director of the IAC. All this has been described earlier. What a cross to have had to bear!

INVESTIGACIÓN ASTROFÍSICA

física solar

física estelar e interestelar

formación y evolución de galaxias

exoplanetas y sistema solar

cosmología y astropartículas

instrumentación astrofísica

Fig. 16.1 Some of the IAC's fields of astrophysical research. (Credit: Inés Bonet/IAC)

This critical spirit that I have been describing materialized in other directions. Critical comparative studies of the IAC's publication record have been made from time to time, and on many occasions specific internal working groups have been tasked with reviewing the IAC's research in comparison with other leading centres of the world. I recall, for example, a 'strategic group for quality in research' being set up in 2001. The group presented a complete 22-page report to the Governing Council. Many of its judgements were more critical than those issued by the various external assessment committees.

To become world class, a research centre must be capable of attracting and retaining the best. To attain that goal the centre must already have very good researchers, top-class astrophysicists with an appetite for hard work. This is beyond doubt the best magnet for talented young scientists. But in this capitalist world keeping the best is intimately linked to what we are able to pay them. Bearing in mind the enormous disparity, in the world of football contracts are directly linked to the pay and other facilities, and tax advantages on offer. Obviously, the IAC could never compete with the well-heeled first league centres of research. All that remains to us are complementary incentives that are also important for a vocational research worker. In our case there are our famed Observatories, replete with advanced instrumentation, and the opportunity to work in an attractive international scientific environment, together with the Canarian climate and quality of life.

Fostering contacts between young researchers and the best in their fields is particularly important for completing the training of researchers and keeping them competitive. And it has been our policy from the beginning to send our people to other centres and invite distinguished scientists to the IAC. Depending on the resources available at any given time, we have made use of programmes of grants, sabbaticals, etc. When the Institute was built on its present site in the eighties, six comfortable rooms were prepared to enable important astrophysicists coming to the Canaries to observe to stay with us for a few days. These rooms served very well in promoting direct contact with our young researchers. Quite a few of the fruitful collaborations of the IAC were born of these encounters. In the same spirit, good use has been made of the Winter Schools (described in Chap. 19), which continue to provide an opportunity for our young doctoral students to get to know and interact with the best and to delve into the most burning issues in astrophysics that are covered each year.

In 1995 the IAC, with funding from the BBV Foundation, organized a very special international colloquium. The great masters of astrophysics and our young researchers gathered to examine still unresolved key astronomical problems. For a week, a handful of students had the good fortune to lodge at Hotel Botánico in Puerto de la Cruz with G. Burbidge, E. M. Burbidge, M. S. Longair, D. Lynden-Bell, G. Münch, I. Novikov, D. Osterbrock, B. E. J. Pagel, M. J. Rees, A. Sandage, and R. Sunyaev.

Cambridge University Press published the encounter under the title *The Universe at Large (Key Issues in Astronomy and Cosmology)*, and the IAC produced videos of the discussions. This book was a milestone, as some of the best astrophysicists of the time, whose influence has spread beyond the IAC, have commented to me. It was an off-the-beaten-track review of our knowledge of the Universe that is still relevant today and packed with suggestions for present and future research.

As an organization grows and its activities diversify, it is easy for individuals to confuse their own tasks with the aims of the organization. At the IAC, to underline that research is the fundamental task, we have repeatedly made it known that, except for researchers, all other personnel, the Director included, were 'support personnel'. All are necessary and indispensable, but strictly in pursuance of quality astrophysical research. Observatories, technology, management, and so forth are not aims in themselves, but tools for doing research and science outreach.

A research centre must have an advanced management system. We have therefore striven, within the tight constraints imposed on a Spanish public entity subject to a thousand bureaucracies, to equip ourselves with a modern

nature
International journal of science

Letter   Published: 13 February 1992

A 6.5-day periodicity in the recurrent nova V404 Cygni implying the presence of a black hole

J. Casares, P. A. Charles & T. Naylor

Nature **355**, 614–617 (1992)   Download Citation ±

Abstract

THE X-ray transient source GS2023 + 338 was discovered in out-burst by the Ginga satellite in 1989 (ref. 1) and has since been identified with the previously known recurrent nova V404 Cygni[2]. This system is recognized to be a low-mass X-ray binary[3], with X-ray behaviour similar to black hole systems[4], but attempts to deduce an orbital period from photometry[5-9] and spectroscopy[10,11] have yielded modulations with periods from 10 minutes to 6 hours. Two years after the outburst, we have used the William Herschel Telescope to find absorption features in V404 Cyg characteristic of a late G or early K star with a radial velocity curve of amplitude 211 ± 4 km s^{-1} and period 6.473 ± 0.001 days. The deduced mass function of 6.26 ± 0.31 M$\odot$ is a firm lower limit to the mass of the compact object, which for reasonable assumptions of orbital inclination and companion star mass must be a black hole with probable mass in the range 8–15.5 M$\odot$. We consider this the most persuasive case yet for the existence of a black hole.

Fig. 16.2 Some examples of outstanding research work carried out at the IAC. *Left:* Confirmation of the existence of a black hole component of the V404 Cygni low-mass X-ray binary system. (Credit: Springer Nature Limited.) *Right:* Identification of a white-dwarf binary as the progenitor of SN 1006. (Credit: Springer Nature Limited)

administration. We have provided the Research Division with a support secretariat for the researchers, and there is even a scientific editor to help them with the preparation of their articles. Nevertheless, our quest for optimization has always been obstructed. Of course, if we compare ourselves with the sprouting bureaucracies of our universities and many other research institutions, we turn out to be fortunate, but that is little consolation.

We have tried many things, more or less successfully, to improve our research. Among these I recall the monthly 'Director's Seminars', in which the scientists would, one by one, talk about what they had been doing. There is also 'Our Science Day', which is celebrated annually, and during which we meet at a local venue, removed from our daily environment to hear what our researchers have been doing during the year. There are, of course, many other activities that it would be prolix to mention here.

One thing that has proved beneficial is the periodic change of the head of research. The incumbent is directly responsible for organizing and managing research at the Institute. Remember that the course of the IAC's research efforts is determined by the Research Assessorial Commission and is sanctioned by the Governing Council. This regular changing of the head of

Fig. 16.3 Leading astronomers who lectured at the 1995 course in Tenerife organized by the IAC. The lectures were gathered in the volume *The Universe at Large (Key Issues in Astronomy and Cosmology)*, published by Cambridge University Press. (Credit: Miguel Briganty/IAC)

research has served to introduce fresh ideas into the division; it also ensures that those researchers who take on this 'social service' do not take time off for their personal research, for the pressure to keep up in the competitive world of research is well known.

I must stress that heads of divisions really must abandon their research owing to the full-time dedication required for the task of organizing things for the benefit of others. They are to be greatly thanked. An extreme case was that of Carlos Martínez, who was the first head of instrumentation and later for many years Deputy Director of the IAC, a permanent support to the Director responsible, among many other things, for taking on the thankless and monotonous, but fundamental, tasks of administration and management. Many thanks, Carlos, from me personally and in the name of the IAC.

Endogamy must always be guarded against. It is all too comfortable to get used to people who work well with you, who did their theses with you, even though they might lack brilliance, people who have devoted years to doing the routine and dull repetitive tasks associated with observational research. It is hard to make way for somebody better from outside, but it has to be done in order to survive in this so competitive and inhumane world.

Gaining tenure is a just aspiration—even more so, and rightly—in these times when job insecurity looms large. Obtaining a fixed position as a civil servant in Spain is a guarantee of a secure future, but such posts are hard to come by because they are few and far between, and the competition is fierce. But once ensconced in a permanent job nobody can fire you, however unproductive you might become. There exists the danger that obtaining tenure encourages settling into an easy life and resting on one's laurels, although this is an infrequent occurrence in research. It can also happen that the person hired as a supposed genius, although a bit odd and extravagant, turns out to be just odd and extravagant. I must confess that the heterodox have always appealed to me because I am convinced that they are necessary for the advancement of knowledge. They are the ones who confront established paradigms with new ideas.

Researchers have a tendency to undervalue, even despise, the tasks of administration and management, which they see as burdensome bureaucracy. There is undoubtedly some truth in this, but it is unavoidable in these times. It is well recognized that in Europe, and in Spain particularly, research centres live on public funding, assigned and invigilated by politicians of the day with their own views and understanding of things. For that reason, the regulations concerning budgetary control employed by Spanish public organisms are not devised to take into account the peculiarities of scientific research. These truisms are deciding factors that force our centres to carry out research that adapts to generic norms, created for circumstantial political reasons that change according to the party in power. An example of this and of the blinkered attitude of our rulers, who still see R&D as a luxury afforded by rich countries rather than as a profitable investment, are the paltry annual budges assigned to research. How is it possible, things being as they are, that any competitive research at all is being done in Spain?

Viewed with the logical reasoning of some of our best researchers, all this is incomprehensible, even more so when they see their treasured and valuable research harmed by mundane considerations. The situation is yet more frustrating for the more fundamentalist ones, who are dazzled by the environment and achievements of the best centres in the world, those with great traditions, abundant funding, and flexible contracts.

I shall end this chapter on an optimistic note. The IAC obtained hard-won accreditation as a 'Severo Ochoa Centre of Excellence' on the creation of that Spanish institution; it has held on to that accreditation ever since. I confess that one of the things I like and enjoy most is to walk through the corridors of the Institute, or enter the cafeteria, to see so many young people, all unknown to me, from different countries and cultures.

I like to think that this ferment of doctoral students from all over the world is a guarantee of the healthy condition of the research that is being done at the IAC. For not just anyone may enter the IAC as a doctoral student. While we continue to attract such prized young astrophysicists, we are doing well.

Fig. 16.4 Stephen Hawking with IAC students in 2014, when he visited Tenerife to participate in STARMUS. (Credit: Luis Chinarro/IAC)

17

Astronomical Instrumentation, Technology Transfer, and Its Impact on Industry (1974–Present)

From the start, both necessity and my personal conviction ensured that the building of instrumentation would be a normal activity for the IAC. I have already told (Chaps. 2, 3, and 6) how, faced with the lack of suitable site-testing instruments, we got down to work in a practical manner, unembarrassed and without complaint, using the tools we had at hand. I was a post-War child, the product of an era in which one made the best of what was available, and in which skill and ingenuity made up for shortages. Furthermore, I had come with my experience at the Daza de Valdés Optics Institute, where we put together our own instruments because of shortages (Chaps. 2, 3, and 6).

When I began to study astronomy and discovered astrophysics, I realized that all the great paradigm-shifting discoveries involved technological leaps forward in scientific instrumentation. It would be necessary to build new instrumentation in Spain too. We had to shake off, once and for all, Unamuno's dictum, 'Let others invent'.

I have begun this chapter by repeating what has always been the philosophy that led the IAC to become a research centre with its own programme of instrumentation development and technology generation. Already in the seventies, motivated by our desire to develop local technology through astrophysical instrumentation, we daringly designed and built what we called the 'automated telescope', conceived for measuring the various components of nocturnal sky brightness. It was a highly automated instrument comprising an array of seven small telescopes, controlled by a NOVA 1200 computer (the first operating computer in the Canaries in fact). The NOVA 1200 marked the birth of the Computer Centre of the IAC, and also that of the University of La Laguna.

© The Author(s), under exclusive license to Springer Nature Switzerland AG 2021
F. Sánchez, *The Rise of Astrophysics in Modern Spain*, Astronomers' Universe,
https://doi.org/10.1007/978-3-030-66426-8_17

Ever since we had confirmed the exceptional astronomical quality of the Canarian skies and realized the possibility of exploiting this natural resource, we began to develop our own home-grown capacity for the development and construction of scientific instrumentation. In a sense, technology is an especially valuable by-product of fundamental research, and we wanted it to be so in Spain as well. I had an uphill struggle with some of the more puristically minded researchers to get them to understand this. The Canarian sky was also put to use in the same way. With national industry in mind, we decided that our Observatories should be not mere observing sites but also generators of new technology.

In 1974, the University Institute of Astrophysics of the University of La Laguna having then been recently created, we set up mechanical workshops in the basement of the Chemistry Tower, and optics and electronics workshops on the fourth floor. That is how we initiated our present optics, mechanics, and electronics capacity to realize our aim of designing and building our own astronomical instrumentation.

Today those early dreams have become a reality. We began by observing with foreign telescopes and other instruments, but we have now developed the capacity to build the biggest and most advanced optical–infrared telescope in the world (see Chap. 21). The Gran Telescopio Canarias (GTC) has a 10.4-m primary mirror. The GTC was designed and developed at the IAC, and was built mostly in Spain. Consequently, the Spanish companies that participated in this venture learned a great deal and have built up a substantial order book for scientific instrumentation. To this we may add another highly significant fact: there are at present two world-class projects to build large 4-m telescopes for solar observing, the American 'Daniel K. Inouye Solar Telescope' (DKIST), to be located in Hawaii, and the 'European Solar Telescope' (EST), destined for the Canaries. Both projects are being headed by IAC astrophysicists: Valentín Martínez (DKIST) and Manuel Collados (EST).

To return to the beginning, we began working on astronomical instrumentation as soon as it became possible, precariously of course, but fully cognizant of our goal, and with will and enthusiasm. When the IAC was born in 1975 and we obtained from the Island Cabildo of Tenerife some land and a number of uralite huts in which to house the IAC's departments, we transferred our primitive Technical Services Department there; it was the forerunner of our present Instrumentation Division.

But before entering into details I shall relate how we chose to get involved with technology. When we were certain that modern telescopes would be brought to the Canaries, there were two opposing tendencies among the few of us that made up the incipient Astrophysics Institute: there were those of us who thought we should make good use of the opportunity to learn to do astrophysics, while at the same time attending to its associated technology; others, however, thought it better not to waste time and resources in developing local technology, but instead concentrate on doing science, and leave the boring and costly business of instrumentation to the telescope owners. We held a meeting to vote on which direction to take (our way of doing things at the time). Those of us who held that it was worthwhile acclimatizing this type of technology in Spain won the vote. This provided us with comprehensible arguments to put to the politicians in order to get the funding needed to proceed.

Rather than weary the reader with a list of 'gadgets' that we have been developing and fabricating up to the present time, I shall simply mention those that were most significant for their difficulty, or for their scientific or technical consequences.

Perhaps the first complex technological project undertaken by the IAC was the instrument we referred to as the 'automatic multichannel telescope for photometry and polarimetry of the zodiacal light and other diffuse sources'. This instrument was a six-channel spectrophotometer that had been devised in the early seventies; it crystallized as a consequence of a research grant that we had obtained for this purpose from the IBM Fund of the Computer Centre of the Complutense University of Madrid. Its development suffered from scant manpower resources and a strong limitation on funding. We obtained small grants from the Juan March Foundation and the National Commission of Space Research (CONIE, after its Spanish initials). We commissioned the mechanics and optics from REOSC (France), but the electronics and software were developed in Spain.

I remember an incident that introduced us to the fierce competition among large companies and their underhand practices. We had to acknowledge that a computer would be needed to govern the automation of the instrument and, on the advice of the Nuclear Energy Council (JEN), we chose NOVA, rather than Hewlett Packard, Honeywell, or IBM—the favoured brands of the moment. We chose NOVA for its advanced configuration and because it was half the cost of the equivalent machines of the other companies. Hewlett Packard mounted a campaign of denigration that obliged us to challenge them to demonstrate before the Academy of Sciences that the NOVA did not suit our purposes, as they had been broadcasting. In the face of this public challenge, the Director General of the company ended up apologizing to our JEN assessors and came to the Canaries to apologize to me in person while at the same time using the opportunity to advertise his company's products.

In the end, this automated telescope never entered into operation; it 'died' during its transfer to the present headquarters of the IAC. Nevertheless, it marked a decisive step forward in our apprenticeship and instilled daring in us when facing up to the challenge of using computers to control scientific instruments. The telescope's NOVA computer was the beginning of the IAC's Computer Centre, as well as being the first computer in the Canaries, as I mentioned earlier.

Between 1975 and 1978 we developed a small spectrophotopolarimetric telescope to observe the upper atmosphere, with a spectroscopic analysis unit based on a circular variable filter covering the visible and near-infrared wavelength ranges out to approximately one micron, with a single-cell photomultiplier. The instrument was built in its entirety in the mechanics and electronics workshops of the IAC. It was operational between 1979 and 1982, and produced data for the doctoral theses of Pedro Álvarez (later to become the Director of GRANTECAN, S. A., the company that built the GTC) and others. Then came the photometers, also for measuring high atmospheric emissions, designed for sounding-rocket payloads, all to be described in the next chapter.

The IAC-80 Telescope deserves special mention, above all for what it represented for the development of the IAC's Instrumentation Division. It was truly a test-bench for the IAC's fledgling engineering, a basic guide to the world of large telescope technology. Work began on it in the eighties, under the supervision of Maximino Galán, but it was not until a decade later that Carlos Martínez succeeded in finishing the instrument, which served as part of the apprenticeship of our young engineers and gave us the confidence to undertake the enormous challenge of the GTC. The IAC-80 has an 83-cm primary mirror, a German equatorial mount, a focal ratio of 11.3, and an effective focal length of 9.02 metres.

The IAC-80 has been in operation at Teide Observatory since 1991; numerous important observations have been made with it and hundreds of articles generated in international journals. Among its most notable results was the discovery of Teide 1, the first known brown dwarf. It was also used to follow the progress of a gravitational lens for ten years to provide information on dark matter in the Universe and obtained images of the celestial body responsible for a violent gamma-ray burst.

The IAC-80 remains in service and is of great value for research programmes that cannot be carried out on large telescopes, and for long-duration observational campaigns. It is also used for Spanish and foreign university student practicals.

Fig. 17.1 The IAC-80 Telescope, designed and built at Teide Observatory by the IAC. (Credit: IAC)

At the end of the eighties, we began the development of systems to obtain images of high spatial resolution through the use of parallel data processors known as transputers. Much was learnt about the development of algorithms for obtaining not only astronomical images of high spatial resolution but also in the development of intensive calculation systems in real time at the telescope. The systems developed were tested on the William Herschel Telescope at Roque de los Muchachos Observatory. The experience acquired using these techniques was applied to an extended source (the sun's surface) by developing a system to correct for first-order atmospheric perturbations (movement of the solar image) on the Vacuum Tower Telescope at Teide Observatory. This system is still operational and has been in daily use on that telescope since then.

The most delicate and complex astronomical instrument yet made by the IAC is the Optical System for Imaging low Resolution Integrated Spectroscopy (OSIRIS) on the GTC. The instrument consists of an imaging camera and an intermediate resolution spectrograph. It possesses various observing modes: direct imaging, long slit, and multi-slit, which may be combined using the 'charge shuffling' facility and tunable filters. For multi-object spectroscopy there is a focal-plane mask designer. OSIRIS entered into service in 2009 and has since then performed excellent observations, giving rise to noteworthy scientific results that have been published in major astrophysical journals. The Principal Investigator of OSIRIS is Jordi Cepa.

Rather than weary the reader further with a detailed list of IAC-made instruments, which those interested may find full descriptions of in the IAC's annual *Memoria*, with OSIRIS I end this description of instrumentation for the visible wavelength range and continue with a brief summary of work done on infrared and radio instruments. I shall then go on to look at the most important aspects of our ceaseless efforts to transfer developed technology to industry.

In 1972 we persuaded Imperial College of Science and Technology, London to set up their 60-inch (1.52-cm) infrared Flux Collector at Teide Observatory. This was an important development for infrared astronomy in our country. The flux collector was later renamed Telescopio Carlos Sánchez in commemoration of our late colleague. With Professor James Ring's innovative design, it was at the time the largest infrared telescope in the world. It had a Dall-Kirkham optical arrangement (ellipsoidal primary and spheroidal secondary) and, far from being a mere 'flux bucket', its optics turned out to be surprisingly good.

As soon as the telescope arrived, we got our people involved in its maintenance and operation to learn these new techniques—a strategy that we always adopted whenever advanced instrumentation arrived at the Observatories (see the next chapter). By the following decade the IAC was able to construct an infrared instrument for the telescope with a linear 32-element infrared detector. Later came a two-dimensional array infrared camera (CAIN) and an infrared photometer (FIN).

The IAC's most significant advance in this wavelength range was the design and construction of a double infrared spectrometer with cryogenically cooled diffraction gratings operating in the wavelength range 2.5–12 μm for ESA's *Infrared Space Observatory* (*ISO*). The instrument was called ISOPHOT-S and will be described in the next chapter.

The most complex and versatile infrared instrument yet built by the IAC is undoubtedly the Multi-object Infrared Spectrograph (EMIR, after its Spanish initials) for the GTC, where it is now in operation. It was designed, fabricated, assembled, and verified in its entirety at the IAC. Its Principal Investigator is Francisco Garzón. The instrument produces wide-field images in the 0.9–2.5-μm wavelength range and also serves as an intermediate resolution multi-object spectrograph. The central feature of the instrument is its 'configurable slit unit', which allows the user to configure and observe in real time with up to 55 slits in the spectroscopic 6.67 arc minute by 4 arc minute field of view. Long slits of different widths can also be configured. The dispersing elements are formed by combining high quality diffraction gratings, fabricated using photo-resistive procedures, with large prisms of high refractive index. Observations can be made with eleven narrow- and wide-band filters. Working in this wavelength range of the spectrum requires the detectors to be cooled in a cryostat to 77 K (-196 °C).

As a result of all this effort, the IAC has gained a high reputation in infrared instrumentation, which is why it is now participating in the design of infrared instrumentation for forthcoming artificial satellites and the extremely large telescopes now being designed.

As more new telescopes and instruments arrived at its Observatories, the IAC followed its strategy of continuing to expand its scientific and technical capacity. The same procedure was always followed: IAC staff would accompany our foreign colleagues in order to learn and adapt these techniques and the associated science to the IAC. The same tactic was employed in the case of radio astronomy.

The Cosmic Microwave Background (CMB) is radiation that reaches us from the first moments of the Universe and has become one of the main topics in the study of the early Universe. Its importance was first realized by

Fig. 17.2 The construction of EMIR (now in operation on the GTC) nearing completion at one of the IAC's workshops. (Credit: Pablo López/IAC)

twentieth-century cosmologists. Photons from the CMB are detected today in the millimetric and centimetric wavelength ranges as background radiation in the microwave region of the spectrum with a temperature of 2.726 ± 0.004 K using radio-astronomical techniques.

At the beginning of the eighties, the United Kingdom's Jodrell Bank Observatory (Nuffield Radio Astronomy Laboratories) told us they were interested in setting up radio telescopes at Teide Observatory because of the site's special atmospheric conditions. We began negotiations and, in our usual style, sent a resident astrophysicist, Jorge Sánchez, to England to carry out research with them and learn about radio astronomy. We signed an agreement with the University of Manchester, as a consequence of which three radiometers, by joint agreement designated the Tenerife Experiment, were set up to measure CMB anisotropies in the frequency range 10–33 GHz, with an angular resolution of 5 degrees. The Cavendish Laboratory of the University of Cambridge joined the agreement, and in 1984 systematic observations of the cosmic microwave background commenced. Jorge Sánchez later moved to solar physics, and Rafael Rebolo took over the project. Since then, Teide Observatory has hosted most of the European terrestrial experiments concerning CMB anisotropies, and Rafael Rebolo is now leading the IAC team in this field.

In 1987 the first results from the Tenerife Experiment, placing more restrictive limits on CMB anisotropy, were published in *Nature*. IAC astrophysicists were among the coauthors. This set of results provided the first independent confirmation of the celebrated discovery of anisotropies in the CMB by the *Cosmic Background Explorer (COBE)* satellite.

Between 1998 and 2008 we took on the challenge of designing and building the COSMOSOMAS (COSMOlogical Structures On Medium Angular Scales) experiment to separate Galactic synchrotron emission from the CMB and microwave emission from free–free interactions in the interstellar medium. With a resolution of 1 degree, the instruments were designed to observe in the frequency range between 10 and 20 GHz. The experiment was located at Teide Observatory and comprised two sets of antennas with cryogenically cooled receivers operating at 11, 13, 15, and 17 GHz. The antennas observed the sky in circular scans obtained using a flat mirror rotating at 60 revolutions per minute. The antennas concentrated the radiation into various receivers housed in cryostats at a temperature of 20 K (–253 °C). Built entirely at the IAC, the telescope produced hundreds of maps which, when combined, reached a sensitivity of 50 μK at 1-degree angular scales at each frequency. Even today, no experiment has surpassed this sensitivity and sky coverage at these frequencies and angular scales.

The main interferometric experiment at Teide Observatory was the Very Small Array (VSA), operational between 2000 and 2008. It was a joint project of the Cavendish Laboratory (University of Cambridge), Jodrell Bank Observatory (University of Manchester), and the Instituto de Astrofísica de Canarias. It comprised fourteen 30-cm antennas, each observing at a frequency of 33 GHz. The receivers used high electron mobility transistors cryogenically cooled to 15 K (-258 °C).

The VSA's objective was to study the structure of the CMB at angular scales from 10 minutes of arc to 2 degrees. It also carried out many observations of the Sunyaev–Zel'dovich effect, which is the inverse Compton interaction between photons from the CMB with the electrons in the plasma in clusters of galaxies.

As usual, we proceeded by acquiring experience in the development of radio-astronomical instrumentation and data analysis techniques that permitted the participation of Spanish scientists and technicians, and firms from our country in international astrophysical projects, both ground- and space-based. I dedicate the next chapter to the development of space astronomy in Spain.

The culmination of the IAC's efforts in radio astronomy has been the QUIJOTE-CMB (Q-U-I JOint TEnerife CMB) experiment, whose scientific purpose is the characterization of the polarization of the CMB and other Galactic and extragalactic emission processes on large scales in the frequency range 10–40 GHz, the ultimate aim of the experiment being to find and constrain evidence of B-mode polarization, and hence of the generation of gravitational waves from the Big Bang. The polarization measurements of QUIJOTE will complement at low frequencies and help better correct for Galactic contamination observations obtained by the *Planck* satellite.

The two now operational telescopes of QUIJOTE are located at Teide Observatory. The IAC is leading the project, with the participation of the Physics Institute of Cantabria (Santander), the Department of Communications Engineering (Santander), the University of Manchester, the University of Cambridge, and the engineering firm IDOM (Bilbao).

The optical system of the experiment consists of two antennas in an optimal configuration to carry out measurements of the polarization of radiation. Each antenna has a 2.25-m primary and a 1.9-m secondary. In the first phase of the experiment the first antenna used two instruments, the first performing simultaneous observations at four frequencies (11.2, 12.9, 16.7, and 18.7 GHz). The second instrument is equipped with 31 polarimeters operating at 30 GHz.

Fig. 17.3 The telescopes of the QUIJOTE experiment at Teide Observatory. (Credit: Daniel López/IAC)

In its second phase (now in progress) the QUIJOTE experiment incorporated an additional telescope, similar in its characteristics with the first, with 31 polarimeters operating at 40 GHz.

I now turn to our sustained efforts to ensure that the technology devised for astrophysical research should find a more general use. It is well known that fundamental, basic—or however you care to term it—scientific research, including astrophysical investigation, generates advanced technological developments that almost immediately find uses in other fields. The first applications of these new developments are usually military in nature, and later find their way into medicine. The best-known example may be the atomic bomb, internet, and various types of scanners. For ethical reason, we include medicine in our technology transfer objectives (possibly also because our armed forces have shown no interest in us). Unamuno's 'Let others invent' persists in our collective subconscious, as demonstrated by the rare use of autochthonous patents.

Let us begin with ophthalmology for its optical affinity and for the common understanding we share with lecturers in this specialty at the newly created Faculty of Medicine of the University of La Laguna. Towards the end of the seventies we jointly developed a reflection spectrophotometer (named the *Espectroftal*) capable of performing spectral analysis of the light reflected by ocular structures and deriving the amount of pigmentation. The *Espectroftal* was awarded the Agustín de Bethencourt Prize for Research of the Caja de Ahorros de Canarias.

Following this line, between 1983 and 1985 we developed a new kind of ophthalmological instrument for the study of the visual field that used a simple cathode ray tube (a television screen) as an element to visualize sight tests. The aim was to provide an economical and simple tool for the early detection of glaucoma. An instrument, the *Hipocampus*, was finally produced, patented, and commercialized. *Hipocampus* was the fruit of a collaboration between the Faculty of Medicine of the University of La Laguna, and MEL, a private Canarian company set up for that purpose.

Later, between 1985 and 1987, the *Oftacron*, another ophthalmological instrument, was developed for spectral observation of the retina. It answered the need for a non-invasive exploratory tool to examine the retina for various illnesses that result in the modification or deterioration of the retina. The

Fig. 17.4 Poster for a meeting ('The Microprocessor and Its Impact on Industry') organized jointly by the IAC and ASINCA (Canarian Industrial Association). (Credit: IAC)

instrument was destined for use as a tool by the researchers of the Ophthalmology Service of the University Hospital.

Between 1988 and 1989 *Texcan* was built. It was a medical image-processing tool, mainly for X-rays and echograms, for use by researchers at the University of La Laguna.

There have also been collaborations with other departments of the University to make significant improvements to marketable instruments and develop new ones. It has been a source of great pride and satisfaction to us that our collaboration with University medical researchers has been so fruitful, giving rise to patents that international companies in the exclusive field of medical instrumentation have used in the instruments listed in their commercial catalogues.

Those Spanish research centres that have tried to transfer technology to industry know full well how difficult it is; apart from any legal obstacles, our mentality and habits often impede the process. At the IAC we have encountered the same problems, but we have occasionally achieved success and are persistent. The most direct way of technology transfer is to hand over, without any fuss, the development and prototype to a commercial company. We chose this means in the seventies, by providing the concept and prototype for a solar parabolic trough to Energía Solar Española S.A. I believe it is still on the market.

The simplest way to transfer technology is to build instruments jointly with an industrial firm. Of course, there being at the time no technological infrastructure, we had to build up our own advanced workshops and laboratories. We also offered our facilities to companies, and our laboratories for calibration and tests.

Yet another variant of technology transfer is to contract out the building of instruments, or parts thereof, to local firms. Eventually, the time arrived when everything that could be made outside the Institute, budget permitting, was outsourced. Naturally, the knowhow and economic returns stay with the firm. Notable examples of the success of this procedure are the international orders received by Spanish firms that participated in our space projects and the building of the Gran Telescopio Canarias.

We have always liked to interpret this as a low-key means for public administrations to stimulate and fund national industry.

We tried on a number of occasions to implement local mechanisms and entities that would allow an increase in the effectiveness of technology transfer, while at the same time bringing some economic return to the IAC for the scientific and technical knowledge transferred, and for our powerful international contacts. So that the reader may gain some idea of the nature these battles, for they have a certain element of interest, I now relate a number of examples.

Fig. 17.5 The *Oftacron*, an optical instrument for examining the retina. Designed and built at the IAC. (Credit: IAC)

Our first attempt along these lines was the creation of Galileo Engineering and Services, S.A. It was a curious and significant story well worth mentioning. Our resounding inaugurations of 1985, with heads of state, Nobel laureates, and ministers from all over Europe, had received wide media coverage that aroused the passing interest of some of the large Spanish banks. The then President of Banco de Bilbao, José Ángel Sánchez Asiaín, ever conscious of the stimulus generated by the use of Spanish technology, well understood my desire that the technology we were generating at the IAC should be transformed into marketable products. He persuaded the President of Iberdrola to participate in the venture to set up a company with this aim, and we saw the convenience of involving a local Canarian entity. Thus, in 1988, was born the Galileo company, in partnership with the Instituto de Astrofísica de Canarias, Iberduero, Banco de Bilbao, and Caja Canarias, thereby uniting scientific, technological, and financial interests. The capital amounted to 120,000,000 pesetas (720,000 euros).

Our first objective was to put on the market an original digital cartography product that made use of an original idea of the Institute that enabled the retrieval of the h-coordinate (altitude) using two-dimensional images from artificial satellites. There were no other products with these photogrammetric capabilities on the market, so we decided to patent it. Our commitment, which we fulfilled, was to create an R&D team in Galileo and concentrate all its activities during the first years on developing and getting the product on the market. An initial success of the team was the development of digital algorithms capable of identifying olive trees in aerial photographs. These algorithms were later used by Galileo Engineering and Services S.A. to carry out work for the European Union and the Spanish firm TRAGSA to identify potential receivers of European grants for the exploitation of olive plantations. A new product for the automatic restitution of maps was created that was eventually used by 40 operators working in two shifts. It had a turnover of 600,000,000 pesetas (3,600,000 euros).

Then what happened? The members of board of directors of Galileo decided to bring in lower-level people obsessed with quick profits. The newcomers decided, in order to get direct funding, 'while R&D projects were under development', to outsource to a company called Inforten Services the assembly and comercialization of personal computers imported from Taiwan, Inforten Services being responsible for developing management software for firms. This change of objectives gave other companies in the sector time to leapfrog ahead and put equivalent products on the market. This change of focus of Galileo, S.A., knelled the demise of our dreams.

The merger of Banco de Bilbao (BB) and Banco de Vizcaya (BV), giving rise to BBV, and the later merger of Iberduero and Hidrola to create IBERDROLA brought about changes in Galileo, which ended up as a small company within the group, the IAC now being absent. In 1996 a purchase agreement was negotiated, the new shareholders now being Control de Gestión, Tracasa (Trabajos Catastrales, S. A.), and a group of Galileo directors and technicians. In 2007 Galileo acquired 100 per cent of Tracasa's shares; the capital was now fully Canarian, and the company dedicated itself to the development and introduction of management systems for public administrations. I am told that in 2016 the company was merged with the Maggioli Group, a leader in consultancy in the public and municipal administration market in Italy, with more than 1400 employees and a turnover of 120,000,000 euros.

What did we at the IAC learn from this experience? Frankly, that it was better to 'stick to what you know best'. To transfer technology, at least here, it was not viable to create a new company; instead, transfer had to be done either through already existing firms in the sectors that interested us, or by creating projects or intermediate transfer entities jointly with them.

We never abandoned the idea of creating something useful for exploiting the technological capabilities of the IAC, thus contributing to the creation of a technological business base in the Canaries. In fact, it was part of our early dreams to exploit the skies above the Canarian summits in order to germinate and consolidate astrophysics in Spain, while at the same time enriching the technological capacity of our country. The minutes of the Governing Council of the IAC and its Board of Directors bear this out.

At the beginning of this century we decided to try again, this time with a better thought out plan. We began by seeking and defining the appropriate kind of entity for putting our ideas into practice in order to launch a project, complete with business plan and all. To brief ourselves on the current situation, we visited regions of Spain that had recently created the most advanced technology companies. We finally contracted Inasmet Tecnalia, a company with wide-ranging experience in such matters in the Basque Country, to draft the project, which we presented to the Governing Council of the IAC in 2006 under the title 'Design and feasibility plan for ASTROTECNIA, a Canarian foundation for technological development'. The creation of the project appeared as an objective in the strategic plan of the Institute for the period 2006–2010.

At about that time the construction of the Gran Telescopio Canarias was nearing completion at Roque de los Muchachos Observatory, and the firms involved showed interest in participating in this technological venture. The prestige attained by the IAC in the development of advanced scientific instrumentation certainly interested the business sector of the country. The IAC had access to the most important international projects in the field of terrestrial and space astrophysics, in addition to a proven track record in the constitution of competitive international consortia.

With all these bright prospects ahead there began a tug-of-war of indecisiveness between the central Government and the Autonomous Community of the Canaries, with much haggling over who should contribute how much. It must be remembered that the President of the Canarian Government and minister with the research portfolio are both members of the Governing Council of the IAC. The delays gave rise to differences between the two governments and we were back at square one.

In 2008 a contract was awarded to Inasmet-Tecnalia for the drafting of the business plan with a view to the preparation of a document incorporating the totality of the project. The document served two purposes: first it was to be used as a demonstration of the project for the public administrations and the business sector, and second it was to serve as a base for defining the initial activities and objectives of this new technological space that we wished to create in the Canaries.

IACTEC, as this entity is now named, is conceived as a 'space for technological and business cooperation for the development of advanced instrumentation in the Canaries'. Its business plan was satisfactorily concluded at the end of 2009. In the years that followed, with the economic crisis, the process of launching the project was again delayed. At long last, in 2015 work began on the building of IACTEC headquarters in the Science and Technology Pole of Tenerife (INTECH), located in La Laguna, not far from IAC headquarters. Funding has also been acquired for the hiring of staff.

Finding a location for IACTEC was also a story in itself. Our first idea was to include the site as an extension of IAC premises. That turned out not to be possible because the offer of space by the Canarian Government was withdrawn following a change in government. The Tenerife Cabildo later offered us land close to Los Rodeos Airport, right under the approach path. But the noise would have been intolerable, so we declined the offer. The Cabildo then offered to locate us in a technology park they were planning to build close to the south motorway: also unsuitable. It was we who finally found a space owned by the University of La Laguna that had been abandoned and turned into an illegal squat. Finally, after laborious negotiations with the University, La Laguna Town Hall, and the Cabildo, European funds are now being used for the construction of a Science and Technology Pole, linked to the University and the Instituto de Astrofísica. It seems that nothing serious is ever achieved in a straightforward manner in this country.

Fig. 17.6   Façade of the headquarters of IACTEC. (Credit: Inés Bonet/IAC)

The birth and spectacular development of astrophysics in Spain are strongly associated with the IAC, which has wisely used the exceptional quality of the Canarian Observatories to capitalize on the large international community that uses them. It has also strategically extended its activities in a noteworthy programme of technological development, maintaining a presence in the most emblematic astrophysical projects of the day. What the IAC has yet to achieve is the creation of an advanced technology base in the Canaries. We have pinned our hopes on, and made huge sacrifices for, the creation of IACTEC.

Without a doubt, the presence of the IAC in large consortia for the development of advanced scientific instrumentation for ground- and space-based astrophysics, and the environment of international cooperation in which it moves are IACTEC's greatest assets, as well as being a palpable benefit for its partners, both public and private.

To provide a broader picture of what has just been described, I draw the reader's attention to the IAC's presence in such initiatives as the European Optical Infrared Co-ordination Network for Astronomy (OPTICON); the project to build the European Extremely Large Telescope (E-ELT) and its possible future instruments; instrumentation for the Very Large Telescope (VLT) array; the *Herschel*, *Planck*, and *Imax* astronomical satellites; and participation in many other initiatives, such as HELAS, ARENA, SOLAIRE, and CONSTELLATION.

Neither should one forget the construction of the Gran Telescopio Canarias. The prominent involvement of Spanish companies in the venture has increased the confidence of these firms in the IAC. With such a favourable climate, then, let us hope that the IACTEC venture finally pays off.

Many people are now convinced that the Canaries can no longer rely on a socioeconomic model based on the monocultures of tourism and construction. Owing to the fragility of its territory and the potential that the technological world offers, the backing of activities of a distinctly technological character, such as those related to advanced scientific instrumentation, presents an opportunity of success in an ever more globalized world.

IACTEC will centre its activities on the astrophysics, space, and medical markets. The idea is to begin with a working model based on bilateral agreements between interested firms and the IAC for the development of specific joint projects in these fields. The infrastructures and services of IACTEC will be ready to meet the technological needs of such collaborations.

In the medium term, should these bilateral agreements evolve in a stable manner, there will be the possibility of adopting a more robust legal structure to enable IACTEC to streamline and augment the capabilities of the partner entities. Various models have been examined, and that of a foundation would seem to be the one with the most advantages. At the same time this new entity would live alongside the previous model of activity based on bilateral agreements for those entities that prefer that mode.

18

The Rise of Space Astrophysics in Spain (1942–Present)

One of the strengths of the Spanish space industry at the moment lies in the building of payload instrumentation. Our companies, together with our research centres, have built a multitude of systems and subsystems for scientific space instruments, and continue to do so. Back in the sixties, the involvement of the Spanish space industry was almost entirely limited to civil engineering works for launch platforms and tracking stations.

In making specific references to a large number of space instruments, my purpose is to highlight the fact that Spanish astrophysics, born of the modern observatories built on its soil, took a firm decision to venture into space with the country's technological progress always in mind.

In 1942, during the Franco dictatorship, the Air Ministry—of necessity and so as not get left behind—created the National Institute of Aeronautic Technology (INTA). In 1963 INTA, in order to keep up with the times, was renamed the National Institute of Aerospace Technology, and all Spanish space activity was combined under INTA, which, through agreements with the United States, participated in the supervision of NASA space tracking stations located in Maspalomas (Gran Canaria) from 1960, Robledo (Madrid) from 1964, Cebreros (Avila) from 1966, and Fresnedillas (Madrid) from 1966. There was also an Hispano–French tracking station in operation at Sardina del Sur in Gran Canaria from 1964 to 1976.

From 1966 INTA had a launching site for atmospheric sounding rockets in Arenosillo (Huelva), established with much help from NASA, and from 1959 the naval San Fernando Observatory had a Baker–Nunn camera for the optical tracking of artificial satellites, also as part of a collaboration with NASA. All

F. Sánchez, *The Rise of Astrophysics in Modern Spain*, Astronomers' Universe, https://doi.org/10.1007/978-3-030-66426-8_18

this, naturally, was in the context of the strategic importance of Spain for American interests during the Cold War.

Europe formally initiated its joint space activities by establishing the European Space Research Organisation (ESRO), which Spain joined in 1962. Concomitantly, our country created the National Commission for Space Research (CONIE) in 1963. In 1975, ESRO would become the European Space Agency (ESA).

In the early seventies a high-ranking Air Force general visited the Izaña Meteorological Observatory (then still part of the Meteorological Service of the Air Ministry). I took the opportunity to tell him about my plans for the development of astrophysics in Spain. He was also informed that in 1972 ESRO had appointed me as an expert on its Solar System Working Group. With all the enthusiasm of Franco's engineers for incorporating scientists into the general project of national reconstruction (as Lino Caprubí tells us in his book *Engineers and the Making of the Franco Regime*), he told me that I had to be a member of CONIE. He pulled a few strings, and I was appointed by a Ministerial Order of 17 March 1976 of the Air Ministry a voting member of the commission, which formally comprised military officers.

At the CONIE meetings I began to find out a number of things about the small Spanish world of space—not only about its deficiencies, but also its opportunities. The overriding interest of INTA at the time was launchers and rockets. It was, after all, an Air Ministry institution. And in 1974 it launched *INTASAT* (the first Spanish satellite) with a lot of help from NASA. However, what was noticeable was CONIE's lack of interest in the more scientific aspects, particularly space astronomy. Its priorities were to operate the Arenosillo base, design and develop rockets, launch the first Spanish satellite, build space platforms, and improve existing terrestrial facilities. I could see that they needed research projects that included the development of scientific instrumentation and I set about devising plans to go into space.

In the Canaries we had already opted for getting involved in astronomical instrumentation, so it seemed logical and achievable to us to incorporate space into our plans. To be sure, it was the perfect moment to open up this field in Spain. It was nothing less than a matter of uniting capabilities with determination to tackle something simple that would help us gain experience and knowledge.

At the Arenosillo base they were testing small rockets, some of them built by INTA, and scientific payloads developed by European research centres from Germany, France, Italy, and Great Britain, but none from Spain.

For its part, the IAC's Nocturnal Sky team had developed instruments to observe high altitude atmospheric emissions, and Dr. Eduardo Battaner in

Granada had done his thesis and been working on these topics. It was high time to leave the basement and get into space.

It was easy to persuade my friend Eduardo and the Instituto de Astrofísica de Andalucía (IAA) to embark jointly with the IAC on a research project that included the construction of simple payloads. These would be placed in the empty nosecones of the rockets launched by INTA. The funding required could be obtained from CONIE.

We took up the challenge of designing and building compact instrumentation capable of surviving launch and equipped with data telemetry systems. Battaner supplied the scientific content, and the instruments were built jointly by the IAA and IAC. Our scientific objective was to measure the luminescence at altitudes ranging from 80 to 100 kilometres. A series of mini-photometers (FOCCAs) were built and successfully launched from Arenosillo between 1981 and 1982. J. J. López, S. Vidal, and L. Costillo of the IAA, and Maximino Galán of the IAC were responsible for the construction of the mini-photometers.

We learned a great deal and this encouraged us to take further steps along this path. The difficulty lay not in producing scientific ideas for experimentation from space but in designing and building suitable instrumentation to be carried there.

Fig. 18.1 Mounting the scientific payload, built by the IAC and IAA, on a sounding rocket. This was the start of instrumentation fabrication for space in Spain. (Credit: IAC)

Our next step was more of a giant leap: to face the challenge of designing and building an instrument to be carried on board a space satellite, and not just any satellite. The *Infrared Space Observatory* (*ISO*) posed severe technical challenges: the satellite's instruments had to operate under extreme cryogenic conditions, the cryogen being liquid helium. It was the most advanced infrared satellite of its time, in which NASA and the Japanese Space Agency worked in collaboration with ESA.

Our daring had a solid base and was not merely bravado. When, in 1972, the 1.52-m Infrared Flux Collector, owned by Imperial College of Science and Technology, entered into service at Teide Observatory, we formed a small team of infrared astronomers and technologists. In our negotiations with the British, we laid down the condition that they help us to build a local infrared team. During the testing of the Izaña site for infrared observing, the young Carlos Sánchez was attached to the British team and went to England to complete his training in this new branch of astronomy. He was to become the founder of infrared astronomy in Spain.

With the experience of the 1.52-m infrared telescope behind them, Doctors Mike Selby from Imperial College and Carlos Sánchez from the IAC took up the gauntlet to design a cryogenically cooled dual-grating spectrophotometer for *ISO* to perform simultaneous observations in the wavelength ranges 2.5–5 μm and 6–12 μm. Much to our surprise and joy, the project was readily accepted by ESA under the name ISOPHOT-S, to be incorporated into the ISOPHOT package, which consisted of ISOPHOT-S, two imaging cameras (ISOPHOT-C) operating in the wavelength range 50–240 μm, and a photopolarimeter (ISOPHOT-P) operating in the range 3–110 μm. The Principal Investigator for the entire ISOPHOT experiment was D. Lemke of the Max Planck Institute for Astronomy in Heidelberg.

We began the development of the instrument at the IAC under the leadership of the now Professor Carlos Sánchez, but after his sudden death in June 1985 we felt that we could no longer continue with this responsibility; we nevertheless looked for ways to continue the project in Spain. We finally persuaded the IAA to take up the challenge of the development, fabrication, testing, and supply of various models for ISOPHOT-S in collaboration with the Spanish firm CASA. The IAA did not succeed in carrying the project forward, so the IAC decided in 1988 to resume work on it under the leadership of Dr. F. Garzón. We continued with CASA as principal contractor and worked together with them.

I should add that the IAA has now entered into the space arena, and with vision and great effort have positioned themselves among the European scientific institutes capable of leading space missions.

To return to ISOPHOT-S, the IAC designed, built, and optoelectronically calibrated the instrument's various prototypes that would go to form part of the flight models and replacement parts of the experiment. It should be underlined that calibration in the 2.5–12-μm range had never been done before in our country. We learned much and surmounted endless difficulties, not the least of which was obtaining funding.

For its simplicity and robustness, ESA considered our ISOPHOT-S a key instrument for the mission, and one of its prototypes is exhibited at the Munich Science Museum. *ISO* was launched in 1995 on an Ariane 44P rocket from Kourou, and the mission was a complete success. Its data still continue to be used by researchers today. The satellite fell silent when its helium coolant ran out in 1998.

The Spanish research institutes and companies that participated in the experiment gained knowledge and kudos that have enabled them to be incorporated with greater ease into subsequent space missions.

Shortly after, the IAC became involved in further space projects, a case in point being the *Solar and Heliospheric Observatory (SOHO)* mission, whose objective was to observe above earth's the atmosphere in order to tackle problems concerning the internal structure of the sun, its chromosphere and corona, and other aspects of the heliosphere and cosmic plasma. This satellite was a joint project of ESA and NASA.

As with *ISO*, we based our decision to become involved on our previous scientific and technical experience in solar studies through well-established collaborations with leading institutions in the subject. When the University of Birmingham expressed its interest in setting up its new instruments at Teide Observatory, we followed our normal procedure of sending a promising young student to Birmingham. The student's name was Teodoro Roca, who would become our leading expert in the new technique of resonant scattering. The new instruments were installed at our Observatory. The new discipline of helioseismology was begun, and the IAC was present at its birth.

Helioseismology is a technique that allows us to measure, with great precision, the periods and amplitudes of solar oscillations. These are caused by pressure waves that break out on the sun's surface, their amplitudes being magnified by constructive interference, in the same way that pressure waves produced by seismic events occur on Earth. Through these oscillations we are able to discover the structure of the interiors of both bodies. Just as the study of such waves inside our planet is known as seismology, so for the sun it has been dubbed helioseimology.

Of the eleven instruments carried on board *SOHO*, we were involved in a limited way with only two. One of them was the Global Oscillations at Low

Fig. 18.2 The IAC has participated in the building of scientific instrumentation for space missions, including ISOPHOT-S for *ISO* (*left*), and GOLF and VIRGO for *SOHO* (*right*). (Credits: ESA and NASA)

Frequencies (GOLF) instrument, a resonant-scattering spectrophotometer capable of detecting oscillations with frequencies between 10^{-7} and 10^{-2} Hz. It was led by the Institut d'Astrophysique Spatiale of France, A. Gabriel being its Principal Investigator. Besides the IAC, French teams from the Service d'Astrophysique/Centre d'Énergie Atomique, and the Universities of Nice and Bordeaux took part in the project.

Spain's contribution lay in the polarization subsystem, the polarization mechanisms and their electronic control (CASA), the mechanical structure (IAC), the entrance subsystem, with the door mechanism and its electronic control system (CASA), and mechanical structure (IAC).

The other instrument in which we participated was VIRGO (Variability of the IRradiance and Global Oscillations), which comprised a series of photometers and radiometers for measuring solar oscillation modes, the so-called p and g modes. The instrument was headed by the World Radiation Centre of Switzerland, which appointed C. Fröhlich as Principal Investigator. Along with the IAC, the Institut Royal Météorologique of Belgium and ESTEC's Solar System Division also participated.

Spain's contribution was limited to the AC/DC converters (CRISA), the ground-based electrical support, hardware, and software (IAC).

Spain's participation in the two instruments was led by the IAC, under the supervision of the now Professor Teodoro Roca. *SOHO* was launched from Cape Canaveral on a Lockheed–Martin Atlas II rocket in December 1995. It had a long and productive working life (it was planned for a three-year mission and lasted more than twenty years).

It is not by chance that the IAA should be the second Spanish research centre to become decidedly involved, and with notable success, in space astrophysics (recall their first steps with the FOCCA photometers and ISOPHOT-S). Given that they had a research team with knowledge and experience in high-altitude atmospheric studies, they began to join projects on planetary atmospheres and then went on to many other aspects of planetary astrophysics from space. They are now an established authority in these fields.

Spain's initiation into space astronomy, then, commenced first, as we have seen, with scientific payloads being mounted on sounding rockets built by INTA, and later with the building of instrument components for scientific satellites. The IAC's participation in the *Herschel* and *Planck* missions marked a qualitative leap forward; for the first time Spain had developed software for space missions. Right from the start research centres in Spain preferred collaboration between research teams and industry, an approach that has produced excellent results.

Now, in the twenty-first century, our country continues to participate in such experiments, but this time also building and heading complete instruments, such as INTA's meteorological station on NASA's two Mars rovers. The first Raman spectrometer to be used in space (RLS-ExoMars 18) will also be built in Spain by the University of Valladolid and INTA. It is also the first time that a principal investigator will be a Spaniard (from the University of Alcalá de Henares).

Special mention must be made of Spain's participation in the European Space Agency's *Planck* satellite, which was launched on 14 May 2009. There was a significant participation in the mission by the Institute of Physics of Cantabria, the University of Granada, and the IAC. That participation consisted essentially of two instruments, the Low Frequency Instrument and the High Frequency Instrument, both fed by a pair of antennas that form what may be called the telescope. The spacecraft also had a service module containing all the subsystems necessary to control the satellite and its communications with Earth, as well as the electronics and computers of the scientific instruments. The satellite rotated to produce complete coverage of the sky every six months, with a greater density of observations in the poles of the ecliptic.

The IAC team took charge of providing the system controlling the digital processing of the data, the Radiometer Electronics Box Assembly (REBA), and the on-board software, as well as actively contributing to the design of the instrument's receivers.

It is pleasing to see how space research and technology have flourished and grown in our country. There are research centres and firms, scattered throughout our geography, taking part with recognized efficiency and reputation in an endless number of space projects and devices from various countries. Today we have a foot in the door of all the missions of ESA and in others of NASA, Roscosmos, JAXA, and the Chinese Space Agency. Examples of IAC participations include the scientific exploitation of the *Herschel* and *Planck* satellites; the search for dark matter with BOSS (SDSS-III) and e-BOSS (SDSS-IV);

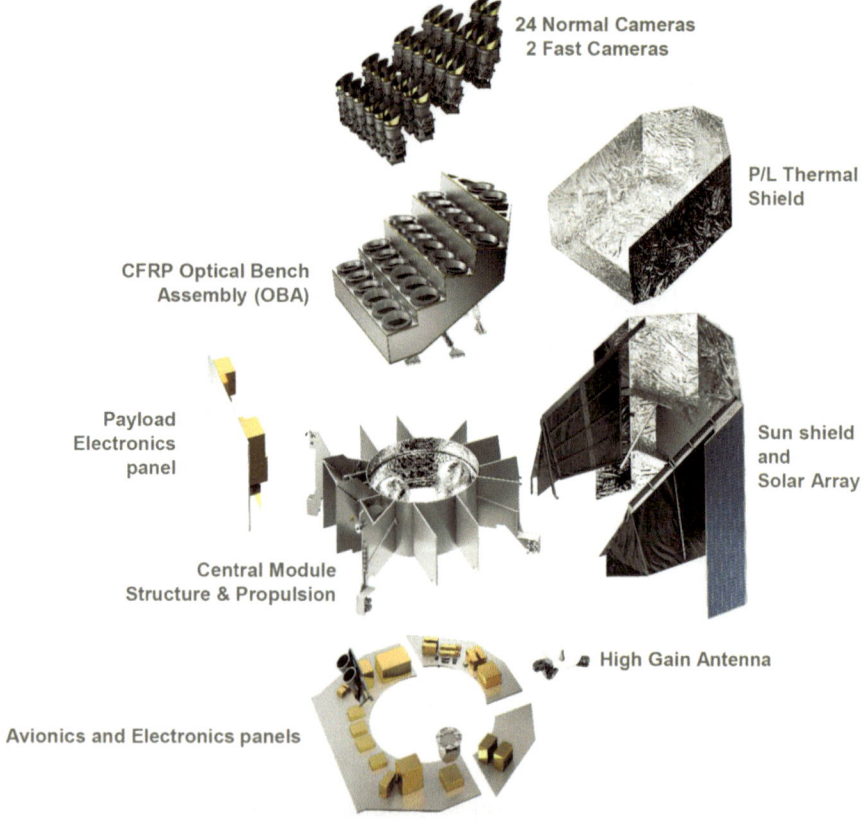

Fig. 18.3 The IAC participated in the development of the control electronics and power supply for the telescopes of the *Plato* mission, whose main scientific objective is the search for and characterization of extrasolar planetary systems. (Credit: IAC)

Fig. 18.4 The IAC contributed to the development of the Control Unit of the NISP instrument, a near-infrared spectrograph and photometer, for ESA's *EUCLID* mission, who purpose is to map the geometry of dark matter in the Universe. NISP will provide photometry, spectra, and redshifts for millions of galaxies. (Credit: ESA)

exoplanets within the framework of ESPRESSO (Echelle SPectrograph for Rocky Exoplanets and Stable Spectroscopic Observations) and *CoRoT*; and the space missions *CHEOPS* and *Plato*.

For me, the most important thing is seeing how our astrophysicists use data obtained from both terrestrial and space observatories in their research, and how our industry is capable of building and supplying space instrumentation.

19

Training Future Generations: The Canary Islands Winter School of Astrophysics (1989–Present)

As I have mentioned previously, one of our basic strategic axes from the beginning of astrophysics in the Canaries has been to provide astrophysicists and technologists with training and skills—a key feature that has permitted us to exploit efficiently the natural resource of the summits of these islands. What is most pleasing is that this drive to educate has progressively extended to all of Spain, producing benefits well known to all, as described in Chaps. 8 and 24.

Let us turn now to a somewhat different training initiative: an astrophysics school different from other well-known summer schools. The Canary Islands Winter School of Astrophysics has a number of peculiarities that render it very attractive. This annual event for young astrophysicists worldwide is now a tradition looked forward to each winter by the international astronomical community. Begun in 1989, it has been a lasting success and has, at the time of writing, reached its thirty-first anniversary. Its fame has permitted us to invite as lecturers the best experts in every field and, because of the intense competition among potential students, only the best are invited to attend. Some of the students of the earlier schools have returned as lecturers in later years. We can claim with pride that several of the most distinguished of today's astrophysicists have passed through this winter school.

This annual winter school, organized by the IAC, is aimed at doctoral students and recent PhDs from all over the world. An attempt is made to bring together the best young researchers and the best scientific authorities to tackle a burning topic in present-day astrophysics. All those attending, both students and lecturers, are accommodated in the same hotel for two weeks in

© The Author(s), under exclusive license to Springer Nature Switzerland AG 2021
F. Sánchez, *The Rise of Astrophysics in Modern Spain*, Astronomers' Universe,
https://doi.org/10.1007/978-3-030-66426-8_19

Fig. 19.1 Participants in the first Canary Islands Winter School of Astrophysics ('Solar Observations: Techniques and Interpretation'), celebrated in 1989. (Credit: IAC)

order to encourage them to interact with one another. We try to get discussion flowing on the topic of the year and contacts for collaborations established.

The idea for the school evolved gradually. In 1979 we organized a School for Young Astronomers, with patronage from UNESCO and the International Astronomical Union. This international school went very well, and we could see how attractive the Canaries and the Observatories might be for young astrophysicists. The schools were also a useful way to learn and make oneself known.

At the beginning of the eighties, with the Treaty for Cooperation in Astrophysics duly signed, the Law of Astrophysics passed, and the IAC and its Observatories inaugurated with full pomp and circumstance, we felt able to take on ambitious projects by ourselves. We were convinced of the importance of training astrophysicists, so we set about designing a different kind of school that we hoped would be better than the usual fare. The key concepts were:

- Interaction between lecturers and students
- The selection each year of a theme that was currently in the limelight
- Holding the school at the beginning of December to take advantage of the international offer of the Canaries in winter time and the low occupancy of hotels before the Christmas holidays

- To try to be different from well-established summer schools while learning from their successes
- To make the IAC and its Observatories known to astrophysicists of the future

With these baselines we began by studying the lay of the land and building on the experience of previous years. As a matter of fact, when preparing each new school, we get the organizers of the previous school together with those organizing the next one in order carry out a detailed critique of the last school with a view to assessing what improvements might be introduced. With this procedure, we maintain a permanently updated protocol of action.

We soon began to make progress; for example, by having the lectures published by Cambridge University Press, a leading publisher in the scientific world. These books form a series that constitutes an up-to-date panorama of topics of the day imparted by leaders in the field. The books occupy a place on the shelves of all specialized libraries. So as to keep the topics as fresh as possible, lecturers are urged to come to the school with their chapters already drafted so that we can get the edited proceedings manuscript to the publisher as quickly as possible.

With a view to fostering contacts, we have always tried to organize matters to ensure strong interaction between lecturers and students. Students are encouraged to make presentations of their research by giving a brief talk or a poster. The lecturers discuss these presentations during a session dedicated to the students' work.

Invariably included in the programme are visits to IAC headquarters, Teide Observatory, and Roque de los Muchachos Observatory. It is tight schedule, but time is set aside, apart from the traditional closing dinner, for relaxation and the opportunity to sample the delights of the Canaries. Our own doctoral students, who are invited free of charge to the school, act as hosts. We want the impression the students carry away with them to be a lasting one for both the work accomplished and the friendly atmosphere.

We take every opportunity to acquaint the participants with cultural aspects of Canarian society. Also, one of the school lecturers is invited to give a public talk at the Tenerife Museum of Science and the Cosmos in La Laguna. We try to promote the Canaries with visits to typical local venues, such as wine cellars, and various social events.

One of our aims is to get students from developed countries to intermingle with those from countries where research is less advanced, and there have been cases in which research collaborations have arisen from such encounters. Roughly half of the students are given grants. The winter schools are

Fig. 19.2 Participants in the twenty-ninth Canary Islands Winter School of Astrophysics ('Applications of Radiative Transport to Stellar and Planetary Atmospheres') in 2017. (Credit: Miguel Briganty/IAC)

expensive to organize, and it is not easy to acquire the necessary funding. In the course of its existence the school has obtained funds from, for example, the European Union, various ministerial departments, Menéndez y Pelayo University, the Cabildos of La Palma and Tenerife, and town halls. As always happens, consolidating these winter schools has taken the time and effort of many people.

By 2018 thirty winter schools had been held. The total participation consisted of 222 lecturers and 1727 students from 30 countries, 630 of those with grants.

Fig. 19.3 Poster for the thirtieth Canary Islands Winter School of Astrophysics ('Big Data Analysis in Astronomy'), celebrated in 2018. (Credit: Gabriel Pérez/IAC)

20

Bringing Astronomy to the Public (1986–Present)

Astronomy outreach has always formed part of our activities since the days when we were still the University Institute of Astrophysics and could organize independently and plan as a body. There were two reasons for this, one basic and the other practical. Being conscious of the attraction of everything relating to the Universe—so intimately connected with wanting to know who we are, where we came from, where we are headed, whether we are alone—we understood that we had to exploit this sense of awe in order to spread scientific knowledge in popular culture, so polluted with astrology and other pseudoscience. Rather than wait for knowledge to filter through slowly in school textbooks, we also felt obliged to report in a direct form to the public that was paying us for the discoveries that we were making. At the same time, we wanted to make ourselves known to the public so that it could see what purpose was being served by all those domes that were sprouting like white mushrooms on the mountaintops of the Canaries.

For all these reasons, from the start we held regular open days at the Observatories, gained media coverage, gave talks, and made ourselves available on request. Since there were so few of us at the time, we took weekly turns to act as ciceroni on visits to the Observatories. It took many years for researchers to be supplanted by specially trained guides.

I always saw an additional social benefit in this. It was important that it should be Spaniards, preferably Canarians, who received the public at the Observatories rather than foreigners operating those enormous and complex telescopes in order to show that we too were people capable of discovering the celestial marvels about which they had heard so much. In that way we would we help our youngsters to break down our ancestral inferiority complex. They

F. Sánchez, *The Rise of Astrophysics in Modern Spain*, Astronomers' Universe, https://doi.org/10.1007/978-3-030-66426-8_20

might then say to themselves, 'If these people who are just like me can do it, then why not me?' I believe that in this way we have helped overcome self-imposed limitations so that new generations might fight for their dreams; we have injected self-confidence into the island community. It cannot be quantified but it is positive.

The IAC has always counted among its aims the diffusion of science and technology in popular culture, and was the first Spanish research centre to set up within its directorate a unit specifically for that purpose. It is known today as the Press and Outreach Unit (UC3, after its Spanish initials).

This structure was organized and consolidated in the early eighties by Luis Martínez Sáez. I knew Luis well, we had been colleagues at the Faculty. Intelligent and able, he has great experience in dealing with people and excels in management, organization, and communication. Employed as Head of the Director's Support Team, his many tasks included a mission to set up the mechanisms of science communication as part of IAC activity.

Being so far from mainland Europe, the Canaries require specific strategies. The fruits of these strategies have proved to be tangible: the popular image of the Canary Islands today goes hand in hand with their skies and astrophysics, and astronomical tourism is on the rise. The islanders take pride in their Institute of Astrophysics. The IAC has become a landmark in science and technology, not only in Spain, and the Observatories have come to occupy a key position in European Big Science. Luis Martínez is an authority in the communication of science, having published on the subject and given university courses.

It has been a struggle to get this far and has required continued intelligent effort in the face of incomprehension, even internally. Our own researchers initially saw outreach as a waste of time, but today many of them most willingly dedicate part of their time to it.

The IAC's main thrust in popularizing astronomy and interacting with the public has been the opening up of its Observatories to public visits. Some time ago an empty telescope dome at Teide Observatory was converted into a visitors' centre, and a science cultural park is now being finished (at long last!) next to Roque de los Muchachos Observatory. Each year thousands of people, from school children to international conference participants, film crews, and other groups, visit the IAC and its Observatories. International celebrities from the worlds of science (Nobel laureates included), politics (including heads of state), finance, literature, and art—even show business—have passed through our Observatories. The Canarian Observatories have become an extremely attractive and exotic curiosity. For this reason, our open days also produce a massive turnout.

As examples, I shall review here only a few of our noteworthy events and significant achievements.

To mark the last return of Halley's Comet in 1986, the IAC, in collaboration with Santa Cruz Town Hall, and the Caja General de Ahorros de Canarias, organized a Star Party on Las Teresitas beach. On a summer's night the mountain-enclosed beach was peppered with several live music stands and audiovisual projections that alternated with public talks and astronomical observations. We had to entertain, address, and instruct people until just before dawn, which was when Halley began to rise. The entire staff of the IAC intermingled with the public to help identify and observe features in the night sky with binoculars. Amateur astronomers brought their telescopes. The celebration thronged with people of all ages and social classes—a true astrofest!

The event was a resounding success. According to press reports, 60,000 people attended from all over the island, and that figure might have been even greater had a bottleneck not formed at the entrance to the beach. This traffic jam prevented thousands more people from arriving and others from leaving until well into the morning. The poet Rafael Alberti said that he missed his flight back to the mainland because of the traffic jam, but that it was worth the hassle to be able to contemplate Halley once more (he had seen it as a child in 1910).

When at last the comet appeared above the horizon it was frankly disappointing. It was hard to distinguish and looked weak and pallid. People are used to seeing beautiful images of glorious comets with distinct hues and large tails. They are unaware that such images require many hours at the telescope, with cameras equipped with special filters. The same frustration is always felt when gazing at the firmament with the naked eye. But, in spite of everything, we are often told that the Halley Party left lasting memories.

A special milestone on our road to injecting scientific knowledge into popular culture was the creation of the Tenerife Museum of Science and the Cosmos in La Laguna, opposite the IAC's central headquarters. The Museum was inaugurated on 11 May 1993. In that same year, the Prince of Asturias, our Honorary Astrophysicist, visited the Museum, whose inauguration he had been unable to chair owing to other official commitments. The Tenerife Cabildo provided the funding, and the IAC, the scientific input.

As always happens, a series of favourable circumstances had to coincide to make it all possible. The project began to take shape after the 1985 inaugurations (see Chap. 13) under the wing of the preparations for the IAC's participation in Expo '92 in Seville, as will now be described. A key event was Adán Martín's accession to the presidency of the Tenerife Cabildo. Martín was a restless and bright young engineer who knew the IAC well since we had collaborated with him in other projects. A man of his time, he was well-travelled and, having first-hand knowledge of famous science museums worldwide, he wanted to establish their equivalent in his own land.

We sketched out the project together. We wanted something modern, simple, but of good quality. The museum should be a place that would awaken an interest in advanced science and technology, above all in children and youth. It should be an attractive place in which to discover through play, where the key issues of science and inventions in everyday use could be taught without pedantry in a direct and straightforward way.

I immediately thought of our young astrophysicist Ignacio García de la Rosa to launch and manage the project. It was he who had set up the 'Astrophysics in the Canaries' exhibition at our Institute with the theme of the 1985 inaugurations. Interested in outreach, Ignacio had toured all the leading science museums in the world, paying for the journeys from his own pocket.

Fig. 20.1 Poster announcing the IAC's 'Star Party' on Las Teresitas beach. (Credit: IAC)

Fig. 20.2 The radio telescope antenna with the logotype of the IAC at the Seville Universal Exhibition in 1992. The telescope was used to transmit messages into space. (Credit: IAC)

Fig. 20.3 The Science and Cosmos Museum built opposite the Instituto de Astrofísica de Canarias by the Tenerife Cabildo, under the scientific guidance of the IAC. (Credit: IAC)

The contents and form of presentation of the modules had to be deciding issues if one wanted to achieve a modern museum capable of drawing the public to its doors. It was not just a question of knowing what to communicate, but also of designing interactive devices to bring ideas to life. One then had to design and build the exhibits, together with, of course, a building in which to house them. In 1987 I put the challenge to Ignacio García de la Rosa, and he accepted.

Adán, Ignacio, and I held endless meetings to mull over ideas and make decisions on issues that cropped up. We chose a young but well-known architect, who designed the building in the form of a spacious street market with a lot of stalls (the exhibit modules). Its layout in the form of a star stands out among the classical buildings of La Laguna, especially when one flies over it when coming in to land at Los Rodeos.

The Cabildo earmarked 60,000,000 pesetas (360,000 euros) for the contents, but the IAC had no means of hiring any staff and the Cabildo had no staff suited to museum work. It was all a struggle. However, the situation gradually improved, and the Institute made available its workshops and staff, who worked extra hours without pay.

A signed agreement between the Tenerife Cabildo and the IAC now regulates the running of the Museum. The Cabildo as the owner manages the museum, but the IAC appoints its director and provides scientific and technical orientation. An express condition of the agreement is the periodic change of director to keep the museum moving with the times and avoid stagnation. There have so far been five directors, each one of whom has introduced new and fertile ideas. The Museum is today a point of reference for science museums in Spain, an added tourist attraction for the island, and a bastion of scientific culture.

Every opportunity is taken to popularize science and bring the IAC to the public's attention. I remember when in 1988 I was named a member of the Expert Committee of Seville Expo '92, which was being directed by Severo Ochoa and chaired by Queen Sofía. The Committee's function, they said, was to make recommendations to the general superintendent of Expo '92 concerning the organization, programming of activities, and exhibit contents. It comprised personalities from the worlds of letters, arts, science, and economics. I found myself among such persons of great standing in finance as Paloma O'Shea, Plácido Arango, Jesús de Polanco, Rafael Lapesa, and Iñigo de Oriol. I put the opportunity to good use, and the IAC presented an outstanding exhibit. We had modules in the Pavilion of the Canary Islands and the Pavilion of the Future; there was also a large radio telescope with the IAC insignia painted on the antenna bowl.

A large infrared telescope that could see what could not be seen by the human eye and could be pointed by members of the public drew attention to the Pavilion of the Canary Islands. In the Pavilion of the Future, astronomical objects could be seen in various wavelength ranges of the electromagnetic spectrum. The public could view through windows galaxies in three dimensions, send messages to outer space, and even 'listen in' to the echo of the Big Bang in which the Universe began.

All of this was possible because we were already embarked on the building of the Museum of Science and the Cosmos, and because both Museum and IAC staff, captained by Ignacio García de la Rosa, became heavily involved, residing for long periods in Seville.

Our efforts were crowned by the installation of an 18-m parabolic antenna, covered with a huge IAC insignia, next to the Pavilion of the Future and large models of space rockets. With the antenna, visitors could send messages while receiving a written communication indicating the distance of the object to which their message had been sent and the time it would take for a reply to be received.

The antenna was raised two days before the inauguration of Expo '92, but when the exhibition supervisor and his team saw the enormous IAC insignia that covered it completely, they raised a storm. It turned out that it was forbidden for shields and anagrams of the participants to exceed two metres in width, and ours measured 18 metres. It had taken an enormous crane and two days' work to set the antenna on its base, and had cost 2,000,000 pesetas (12,000 euros). There was no time to erase the IAC's insignia, and what finally persuaded them to relent was my explanation that the insignia was not propaganda but a simple requirement that all IAC telescopes carry the Institute's insignia. In fact, the telescope remained there for many years, and the IAC insignia could be seen from aeroplanes flying over Seville.

Along the same lines, in 2000 the IAC participated in the Pavilion of Spain at the Hannover Universal Exhibition with a noteworthy exhibit bearing the title 'The Island of La Palma: Look at the Sky and Save the Earth'. This exhibit had clear repercussions on tourist visits to La Palma from Central Europe.

Apart from these grand exhibitions, the IAC has participated in and organized many more, which it would be otiose to itemize here. Suffice it to say that a determination to inform and spread knowledge underlay all these activities. In the same spirit we have organized various competitions for photography, astronomical observing, essays, etc.

But it has not been success all the way. Ever since the 1985 inaugurations we have been trying to arrange for a visitors' centre to be erected at both of the IAC's Observatories; comfortable and pleasant venues in which to relax and take refreshments, while absorbing information concerning the Observatories and the science being done at the telescopes. Unfortunately, in spite of our most strenuous efforts there has been no way to achieve this.

At Teide Observatory we even managed to persuade CajaCanarias to fund an architectural project as a first step towards finding the finances for the building work. But we were never able to get the state, the Canarian Community, or the Cabildo to contribute. In the end we were forced to adapt an empty dome for the purpose.

There was a moment when it seemed that a grand project for a visitors' centre was about to prosper, with telescopes for visitors to carry out their own observations. In order not to disturb the scientific work it was thought that cars and buses could be parked in El Portillo at the entrance to Cañadas del Teide, with shuttles to ferry people back and forth to the Observatory.

Fig. 20.4 Poster summarizing the outreach activities of the IAC. (Credit: Inés Bonet/IAC)

There have been other attempts over the years, but none has succeeded, in spite of the huge numbers of visitors to the Teide National Park, many of whom would like to get to know the Observatory. Now, thanks to the area having been certified as a 'Starlight Reserve', many private companies are providing night-time guided visits to see the stars from designated points in the National Park. With the astronomical tourism boom, the demand is such that many of these companies have degraded the quality of the visits and even the peace of star-filled nights near the Teide. I hope that these nocturnal activities in the Park will soon be regulated.

On the island of La Palma, we still have not managed to get a visitors' centre next to Roque de los Muchachos Observatory. There have been occasions when we were close to achieving it, such as when ICONA signed an agreement of cooperation with the IAC in June 1987, with the full approval of the CCI, for the building, bringing into service, running, and management of an interpreters' centre at Roque de los Muchachos. There was even a national competition, and a project was chosen. Funds were available to carry the project through, but the delays caused by endless bureaucratic red tape and territorial squabbling in La Palma ended with the moneys being withdrawn.

More recently, a private company expressed a willingness to build and exploit what was to be a 'cultural park' through a temporary concession. The Governing Council of the IAC gave its consent, and everything was prepared for launching a call for tenders. But the company had to deal with La Palma (Garafía Town Hall, the Island Cabildo, Taburiente Caldera National Park, the ecologists, etc., etc., etc.). Same story. Another opportunity thrown away.

Changing the subject, in 1986 we employed a journalist, Carmen del Puerto, who today heads the IAC's Press and Outreach Unit. She gained experience in scientific journalism and science outreach, eventually doing her doctorate with me on the interesting topic of how and why such expressions as the 'Big Bang' and the 'God particle', having become popular through media coverage, and can be used as points of departure for science outreach. She initiated, under Luis Martínez's guidance, the *IAC Noticias* magazine, whose mission was to report on the latest news in astrophysics and IAC activities. As was stated in the zeroth issue, the IAC's intention was to have 'walls of glass' so that its achievements would be visible and relayed to people directly. That continues to be our mission today through the use of the latest communications systems, and Carmen continues to bring fresh ideas to the task of outreach.

It has always been clear to us that, if we wanted to popularize science, we needed to rely on the communications media and stimulate journalists to cover science combined with technology. We have always tried to tempt them with attractive news items. For example, in 1987, to mark the inauguration of the 4.2-m William Herschel Telescope at Roque de los Muchachos Observatory, special workshops were organized for the benefit of science journalists. We repeated this format in 1989, when we invited science journalists to the inauguration of the 2.5-m Nordic Optical Telescope at the same observatory.

The IAC has always tried to cater for the information-hungry news media, in the case of television by offering betacam-format images and computer-generated graphics expressly created to illustrate the scientific concepts underlying each news item.

We have always offered future journalists and communicators the possibility of making direct contact with the science and technology being done at the IAC. Students and graduates in journalism, and graduates finishing their master's degree in science communication at Spanish universities, for example, may do their practicals in science journalism at the IAC. Also available are grants and temporary contracts to work at the Institute.

In 1999 Dr Alfred Rosenberg, an astrophysicist, joined the Director's Support Team and was set the task of attending to and clarifying queries from the news media and organizing outreach activities. Rosenberg, as scientific assessor, and other UC3 staff jointly tend to other collectives, such as publishers, institutions, teachers, students, amateur astronomers, and the general public requesting information by letter, email, telephone, or in person. Alfred is an excellent, rigorous, and personable popularizer. He has led highly successful educational projects, such as 'SolarLab' and '100 Square Moons'.

Cosmic phenomena of great repercussion always present a special opportunity, for example the collision of multiple fragments of Comet Schumaker-Levy with Jupiter in June 1994 that had been predicted a year before it occurred. Even though the event was observed with telescopes all over the world, according to the International Astronomical Union, the images taken at the IAC's Teide Observatory were the first to report the collisions. With the use of the EFE Agency, the IAC supplied the news media with daily reports—written, pictorial, and video—during the week of impact. More than 2000 people gathered at the Museum of Science and the Cosmos with a direct link to the Observatory to watch the collisions as they happened during a period of four hours on screens that had been set up for the event.

With the arrival of Comet Hyakutake, the IAC suggested simultaneous observations to amateur groups on the night of Saturday, 23 March 1996. According to information received from some forty groups from all over Spain, over thirty thousand people visited the sites appointed for viewing the comet. The beach at Puertito de Güímar in Tenerife chalked up the highest attendance with 5000 visitors.

Also relating to Hyakutake, the IAC organized a nationwide contest with the title 'The Cometake: Catch it if you Can!' There were three categories dedicated to astronomical photography, children's drawings, and written articles. The contest winners in each category would get a diploma and their work would be published in *IAC Noticias*. The first prize for each category consisted of a three-day visit to the IAC and its Observatories in Tenerife and La Palma. Also included was a night's observing to coincide with the date of closest approach of Comet Hale-Bopp in April 1997.

In 1999, on the occasion of the Leonid meteor shower, the IAC created a project to involve high school pupils from all over the country in observations of the shower. The projects revolved around a 'teaching unit' aimed at teachers of the secondary-level subjects Natural Sciences and Astronomy Workshop. Our main objective was to introduce young students to real scientific research through observation. Meteor storms offer a unique opportunity to bring this kind of experience to fruition: they are spectacular phenomena that require no special observing instruments. This project was the first initiative involving direct cooperation between students and researchers ever to take place in Spain. A total of 1500 students from 76 secondary education centres from all over the country participated.

Building on this experience, the IAC has produced further teaching units. There is a general one on 'Meteor showers' and another on 'Eclipses', which provides instructions for determining the apparent size of the Moon and measuring transit times for the Earth's shadow over lunar craters. Yet another unit, 'Occultations and transits', explains the phenomenology of these astronomical events, and suggests activities and practical work based on these phenomena. I am told the units are very popular.

Primary and secondary school teachers are our best allies in spreading science as an essential part of our culture, and astronomy is a beautiful and effective way of doing this. That is why the IAC maintains permanent links with the educational community, and not merely through teaching material. We give talks at educational centres, and develop teaching projects and specific courses for the training of teachers. In 1990 the IAC, together with the Santa María Foundation, organized an annual course with the award of prizes called 'The Universe and I' for European teachers from educational centres. During the course those attending would receive astronomical training, pedagogical advice, and suggestions for astronomical practical work. The 'Cosmoeduca' project was also set up to help teaching staff develop themes with a scientific, cultural, and human emphasis through the use of astronomical concepts and content.

A continuous effort is made to create teaching courses obtainable from the IAC free of charge. But, apart from aiding teachers, we also extend our educational efforts to amateur astronomical associations and science popularizers; since 2004 the IAC has made available the package 'Space-Time Odyssey', which consists of a series of ten Power Point presentations, each with a minimum of fifty slides with explanatory text, images, diagrams, and animations. It was made to represent a fascinating voyage towards the boundaries of modern astrophysics and is aimed at a wide audience.

Since the eighties, when we began to organize outreach and communication at the IAC, great effort has been expended in publishing, first on paper but gradually going over to digital format to keep up with the times. These publications provide information about the Institute and outreach, taking the form of annual reports, leaflets, posters, postcards, etc.

Since 2002 the IAC has produced two digital magazines: *Caosyciencia*, an interdisciplinary magazine, and *GTCdigital*, which periodically reports on the evolution of the Gran Telescopio Canarias.

The IAC has established a media presence with information, interviews, and articles. For a number of years, starting from 1999, *Canarias innova* was broadcast on Spain's National Radio every Sunday from 12.00 to 13.00 hours and had a wide audience. The programme reported on scientific and technological news generally, astronomical curiosities, and gave updates on both R&D funding and grants.

An activity that is now on the rise is remote observing for outreach. Facilities for such activities have been created at the IAC's Observatories, making use of little-used professional telescopes and robotic telescopes designed for just that purpose. Our aim is to familiarize the educational community with scientific method and practice, while at the same time contributing towards the popularization of astronomy. Also, those persons who have always wanted to operate a telescope but have lacked the necessary instrumentation now have an opportunity to do so through specific programmes.

At Roque de los Muchachos Observatory, the Educational Project with Robotic Telescopes (PETeR), designed for educational centres, has been in operation since 2005. The project makes use of the latest-generation professional robotic Liverpool Telescope, which has a 2-m primary mirror and belongs to Liverpool John Moores University. Part of the observing time on this telescope is managed by the IAC for astronomical outreach projects in Spain.

Another such project is the Open Outreach Telescope (TAD), a system consisting of two telescopes set up at Teide Observatory and remotely controlled via internet using a simple navigating tool. By following a simple protocol, the telescopes may be used as if the observer were there in situ. The telescopes form part of the GLORIA (GLObal Robotic telescope Intelligent Array for e-science) network, which allows free access to seventeen robotic telescopes that continue to be set up worldwide.

There is also TeleAstronomy, a project whose aim is to show every-day activity at a professional astronomical observatory. A number of web cameras located in the control room and dome of the IAC-80 telescope at Teide Observatory allow images and sound to be webcast in real time. The scheme is particularly aimed at pre-university students so that they can see astronomers in action. The idea is for the astronomer on duty—through direct observations, videos, and transparencies—to reveal the secrets of the Universe. Viewers may ask questions and even ask for images to be taken.

Nowadays, since information flows primarily through the internet, this is the main route used by the IAC to persevere in its task of making science a part of general knowledge. For this reason, all such information is provided by the IAC, in both Spanish and English, on its web site (www.iac.es) and on social networks such as Facebook, Twitter, LinkedIn, Youtube, Dailymotion, Vimeo, and FlickrInstagram. The IAC also has its own blog. Among their many other attractive features, these platforms host a gallery of spectacular astronomical outreach images, videos, audios, and CDROMs. In short, the reader wishing to know more about what I have been describing here will find it all on these platforms.

The IAC has received a number of awards for its labours, among them the Special Prize of the Casa de Ciencias (a science museum in La Coruña, Spain) 'for its significant role as a model of how to pursue science without losing sight of the interests of the citizenry, its attention to outreach, and its interest in communicating with society through exhibitions, digital magazines, radio programmes, and many other activities'.

All these activities require hard work, dedication, and vocation. They also require money for both personnel and means, but, sad to say, funding is in short supply owing to the blindness of our politicians in matters of science.

Apart from its meagre budget, the greatest stranglehold on the IAC is, once again, the crushing burden of bureaucracy, a consequence of the redundancies inherent in the Spanish system of political organization. Administrative fragmentation is even more pronounced in the islands, management of affairs being shared among the administrations of town halls, cabildos, the autonomous community, and the state. The duplication and overlapping of regulations have a paralysing effect on everything. 'That many suffer the same thing is but a fool's consolation', as the Spanish saying goes.

In spite of having dedicated an entire chapter to these matters, I have not described anything like the totality of the efforts that the IAC has expended in infusing science into general knowledge through astronomy. My reason for writing this chapter is my firm belief in the importance that these tasks should assume in research centres today. Not so long ago, such chores were viewed as a waste of time, a distraction from the duty of doing science, a mere squandering of resources.

I remember when the University of Padua, where Galileo was teaching for a long time, organized a series of events in that scientist's honour during 1992 on the 350th anniversary of his death. Among these events was a very special conference on the great man to which I had the luck to be invited. In the course of the day, a select group of scientists, many Nobel laureates among them, initiated the sessions by reviewing the many facets of the sage. That night, a renaissance dinner, based on Galileo's own recipes, was served to the accompaniment of music of the period. It was quite a do.

I was particularly impressed with the talk given by Professor Carlo Rubbio, the Director General of CERN, on Galileo as a popularizer. He said, in effect, that the problem that Galileo had with the Church, which in the Middle Ages represented established wisdom, was a consequence of the great scientific leap of the renaissance. It was not so much a religious as a knowledge-based confrontation, a head-on clash between two different ways of seeing the world, a conflict of paradigms.

He used data to contrast the great gulf that has now opened between scientific knowledge, which is increasing exponentially, and pseudoscientific beliefs, which are becoming increasingly popular. This rift, he maintained, is greater today than it was in the renaissance. If Galileo had so many problems just for insisting that the heavens were not perfect, and that the earth orbited the sun, what is going to happen to us scientists today when we even have the temerity to meddle with people's genes? Smiling and gesticulating as only Italians can, Carlo wryly declared, 'They'll burn us! They'll burn us!'

All this was a preamble to an insistence that researchers and their places of work must involve themselves in the popularization of science. Galileo, Rubbio reminded his audience, wrote in Tuscan rather than Latin so that he could be understood. He insisted that we all had both a moral and a pragmatic duty to devote time and resources to explaining in a direct manner the science that we are producing, and that the immense resources needed for Big Science had to be justified before the citizens, who foot the bill.

I was delighted to hear these words from the mouth of this renowned Nobel laureate. It was more or less what we had been thinking all along. It felt like a pat on the back for the strategies we had been following at the IAC from its very beginnings.

A direct recompense for our efforts is to sense the popular acceptance of the IAC in the Canaries in the form of the tangible pride felt by the islanders for their *Astrofísico*. This both encourages and protects us. On more than one occasion popular movements have prevented governments of the day from causing us too much harm, as mentioned in Chap. 12.

I cite two awards by way of illustration. Both awards are in recognition of the research work of the IAC, its efforts in developing technology, its training of personnel, etc., but they are also closely linked to our continuous striving to communicate our activities and science in general.

In 2005 the IAC was awarded the first *El Mundo* Canaries Prize, granted jointly by the Cabildo of Gran Canaria and the newspaper *El Mundo*. As reported in that newspaper, 'Few institutions on the islands have achieved such a status beyond the Archipelago as the Instituto de Astrofísica de Canarias, a body which, from its headquarters in La Laguna and its observing centres on the peaks of La Palma and Tenerife, has obtained the maximum possible yield from the spectacular conditions of the Canarian skies'.

The IAC was awarded the Canaries 2007 Prize by the Canarian Government for 'building fraternity among peoples, contributing to collaborations with other countries, and aiding the progress of humanity'. The prize calls attention to the international visibility that the IAC's scientific activity gives to the Canaries and rewards its great efforts in the popularization of astrophysical topics in Canarian society.

As I said at the beginning of this chapter and underline now, all that has been described here is the work of very many people, and I hope it continues to be so.

We are all aware of the tragedy that is about to befall us with climate change. I am convinced that only if the blight of humanity is capable of immersing itself in a culture of the latest scientific knowledge and of acting accordingly will this world perhaps avoid catastrophe. If not, all that will remain is to await the impact of a giant asteroid so that life can begin all over again on this ill-used planet. Let us enter into reason!

Fig. 20.5 Key speakers at the 2011 STARMUS conference engage in debate under the 10.4-m Gran Telescopio Canarias. *From left to right:* Brian May, Alexei Leonov, Neil Armstrong, Garik Israelian, Leslie Sage (Editor of *Nature*), Francisco Sánchez, Richard Dawkins, Jill Tarter, and Nobel laureates Jack Szostak and George Smoot. (Credit: STARMUS)

21

The Biggest, Most Advanced Telescope in the World (1989–Present)

To simplify matters greatly, we may consider a telescope to be a collector of photons (a funnel that captures and concentrates the light of the stars). The bigger the mouth of the funnel, the more photons it can gather and, therefore, the greater its capacity for capturing light (electromagnetic waves to be exact) from very weak or distant objects. But optics also tells us that the diameter of the mirror determines the resolving power of the telescope. Resolving power is the ability to separate two objects that apparently form a single object to the naked eye. In other words, a bigger diameter enables the Universe to be observed in greater depth and finer detail.

These properties explain the obsession of professional astronomers with telescopes of greater diameter than provided by current technology and their desire to locate them on sites with very special atmospheric conditions in order to ensure the greatest number of nights of high astronomical quality. All this has been explained in earlier chapters.

The atmosphere, always the astronomer's greatest enemy, acts as a thick perturbing medium nuzzled against the aperture of any terrestrial telescope. Clouds prevent the stars from being seen at all, and, even in their absence, atmospheric turbulence smears stellar images. This latter is not so obvious and requires some sort of explanation, which I shall give in an elementary way. We have all had the experience of seeing how objects a long way off on a heated road shimmer, becoming distorted and blurred. The light from distant objects must traverse seething bubbles of hot air, which alter the straight path that the light would otherwise have taken, so that the image on our retina, or any other optical detector, jitters and blurs. This is also the cause of the twinkling of the stars.

© The Author(s), under exclusive license to Springer Nature Switzerland AG 2021
F. Sánchez, *The Rise of Astrophysics in Modern Spain*, Astronomers' Universe,
https://doi.org/10.1007/978-3-030-66426-8_21

The amount of atmospheric turbulence increases proportionally with the diameter of the objective of the instrument with which we observe. Our eye, with its tiny objective, only notices these alterations when the turbulence is very high. However calm and stable the night, we need only look close to the horizon to see how the stars flicker and jitter. If we look to the zenith with a small telescope, we always see the stars twinkling. With a professional telescope several metres in diameter the phenomenon is much more apparent. Only at exceptional sites, such as Roque de los Muchachos Observatory on the highest peak of La Palma, do the stars almost cease to twinkle, jitter about, or merge with one another. Here, the very dark sky is speckled with myriads of bright points of light that fill it completely. It is an impressive spectacle! At sites such as this the atmospheric turbulence is very low and the resolving power of the telescopes located there is greatly improved. That is why today's extremely large telescopes are located at such sites as this.

I thought it necessary to begin by explaining all this so that the need to put telescopes with mirrors greater than 10 m in diameter on the highest peak of La Palma should be understood. There was also an extremely important strategic reason for placing the GTC at the Roque: without a telescope endowed with the latest technology, the Observatory would no longer rank among the top three observatories in the world. Not ranking among the first would lessen the Observatory's ability to attract the ever more powerful and advanced telescopes of the future, necessary for doing competitive astrophysics. Fulfilling the strategic objective of providing the Spanish astrophysical community with a large telescope would also accord with these considerations.

Apart from these powerful reasons, the building of a large Spanish telescope would lever the development of advanced scientific instrumentation, which would produce knock-on effects to boost innovatory technology in the country—one strategic objective more.

It is a source of pride and satisfaction to have achieved these strategic aims. Roque de los Muchachos Observatory continues to be a very attractive site for next-generation telescopes; our scientific competitiveness has greatly increased, and the several dozens of Spanish firms that participated in the construction of the GTC now have bulging order books for the most advanced scientific projects of the day, both ground- and space-based. It is also worth noting that more than 70 per cent of the GTC was built by Spanish companies. That the remainder was not built here is owed to the timidity of Spanish investors, who were unwilling to risk building an optics company in spite of the guaranteed participation of the only firm in the world with experience in this type of segmented mirror (experience gained in the building of the two 10-m Keck telescopes). Every time I think back on this I am filled with sadness and anger.

It is a tragedy that Spain should have thrown away such an opportunity. The French firm SAGEM, which also had had no previous experience with such mirrors, had to be commissioned for the task and is now Europe's leading firm in segmented mirrors, which are now the norm for new terrestrial and space telescopes. It should also be added, to our even greater chagrin, that it was an outstanding Canarian engineer, Javier Castro, designer the GTC's optics, who provided SAGEM with specific solutions to help that company finish the polishing of the mirrors.

I now describe, in broad brush strokes, how this impossible dream of a large Spanish telescope was made possible, bearing in mind our total inexperience in the design and construction of large telescopes and considering that our previous track record consisted of a humble 80-cm telescope (the IAC-80). Another far from trivial concern was the lack of money for such an immense undertaking. There is no denying that it seemed an impossible challenge for our country at the time. In fact, it was the very first Big Science project led and built by Spain, and located on its territory. We had previously always failed to get any Big Science project established in Spain.

We began by exploring the simplest path, which was to work in collaboration with close colleagues, our experienced British partners, who had just built and set up a 4.2-m telescope at the Roque. This Anglo–Dutch instrument, named the William Herschel Telescope, entered into service in 1987 and was the biggest and most advanced telescope in Europe at the time. It had even earned a place in the *Guinness Book of Records*. It was designed by the Royal Greenwich Observatory. So, wasting no time, during the inaugural ceremony of the telescope I made overtures to Alec Boksenberg, the then Director of the RGO. We soon arrived at an agreement on both the desirability of having an 8-m telescope at the ORM and the advantages of building one jointly between the RGO and the IAC. We decided to start work immediately on a pre-project for a telescope of that class; with such a document in our hands, we felt we would have a stronger argument with which to convince our countries to meet the costs and get other countries to join us. To bolster the initiative, we persuaded the CCI (see Chap. 10) to set up a subcommittee specifically tasked with taking a decision on the 'large European telescope'.

We formed a joint technical group headed by Brian Mack (the British engineer of the 4.2-m telescope project) and Pedro Álvarez (head of the IAC's Instrumentation Division). In 1989 we had a written project for an 8-m telescope, but we were able to convince only the British and Spanish authorities of its viability. Spain and the United Kingdom opened negotiations on the basis of a 50 per cent participation by both parties to set up the telescope at the ORM. All went well until, at the eleventh hour, the British were tempted away by the United States to participate in their Gemini Project, which consisted of two twin telescopes to be set up in different hemispheres. The British astronomical community was divided between proponents of the Canarian telescope and those of the American project. I remember the renowned Professor Martin Rees being among the former. Nevertheless, in 1990 the UK's Science and Engineering Research Council opted—a single vote swaying the decision—to join the Gemini Project in preference to the Canarian telescope option. Britain's affinity with the United States and the lesser risk involved tipped the balance in favour of Gemini. After this failure, the Spanish authorities informed us that the matter was closed, and from that moment on the IAC directorate continued alone, both in Spain and abroad, with the idea of a large telescope for the Canaries.

According to legal regulations, the Director of the IAC possesses a number of powers, but the Governing Council is 'the supreme decision-making organ of the Institute', so it was absolutely necessary to get its formal backing, especially for a matter of such magnitude as this large telescope project. As a collegiate body, its decisions had to be taken by persons that formed it at any given moment, and in our case the constitution of the Council depended on the vagaries of the political climate. For this reason, it was necessary to work repeatedly on its ever-changing members, who also brought along with them their distinct political and administrative persuasions. At that time the Governing Council was chaired by the Minister for Education and Science, and the President of the Canarian Government was Vice-chair. The Council was completed by the Sub-secretary of the Presidency of the Government, the Chancellor of the University of La Laguna, the President of CSIC, and the Director of the IAC as voting members, and the Presidents of the Island Cabildos of Tenerife and La Palma as non-voting members. It was extremely difficult to arrive at a consensus, however hard I tried to persuade them that the large telescope was a matter of national, rather than local or party, interest. Round and round went the carousel, dependent on the election results of the day.

Our morale was raised when we obtained an allocation from the European FEDER funds (1994–1999), which enabled us to continue with the pre-design of the telescope. We proceeded immediately with a feasibility study.

I obtained an audience with the President of the Canarian Government, Manuel Hermoso, and with the Economy and Treasury Councillor, José Miguel González, in the small palace then occupied by the Government Presidency in the Plaza de los Patos in Santa Cruz. I have clear recollections of the meeting. Both men were professional engineers, intelligent, with political vision, and sympathetic to the IAC. We got on well with them. Once again, I laid out the whole project to them in great detail, drawing attention to its advantages and difficulties. The project took a decisive step forward when they finally put their full weight behind the project. The Canarian Government would create a limited company of public ownership to give it the required legal and administrative weight and consistency. I am sure that the argument that finally decided them was a quite straightforward one. The numbers I showed them demonstrated how our large telescope would cost more or less the same as adding a 10-m extension to a dock in a Canarian port. The coasts in the islands are so abrupt that Isabel the Catholic once complained, 'It would seem that the ports in the Canaries were built with *maravedís*[1] rather than stones'. I drove home my argument by asking whether it would be more beneficial to the Canaries to lengthen a dock by 10 metres or engage in the construction of a modern Big Science machine, complete with transferrable technology, prestige, and visibility. They saw my point and told me to proceed with the telescope.

After the meeting I perceived that the project had at last begun to take on the appearance of reality and I began to have sleepless nights. The problem that now confronted me was enormous. Only powerful countries with a tradition in Big Science had hitherto taken up the challenge of building an instrument of this complexity and size. This was no mere 80-cm telescope, and it would be far from trivial to acquire the money needed.

I convened a meeting with close and senior members of the IAC and gave them a detailed account of the state of play, voicing my concerns. We formed a working group to face the challenge of this peril-laden opportunity. I was relieved that I would now not be the only one losing sleep.

Once the decision to proceed with such a daring venture had been made, we had to act decisively and smartly on all fronts. We immediately opened up a debate among the IAC astrophysicists on the project to discuss its advisability, opportunities, and risks; for it had to be an enterprise shared by all and not just the directorate. We laid our cards on the table, took due note of the huge risks we were taking, and the great responsibility that the IAC would be assuming. There were wide-ranging discussions lasting several sessions, and—finally—we all agreed to proceed, but taking care to guarantee success and avoid the falling into the trap of the IAC becoming the 'Large Telescope Institute'. We had to maintain the IAC's R&D programme, so the management of the telescope would need to be externalized, but without severing links with the Institute. And that, in effect, is what we did.

---

[1] Maravedís were a variety of gold or silver coins used in Spain from the eleventh to the nineteenth century.

On 11 May 1994, the Canarian Government formally set up the public company GRANTECAN, S.A., and at the first meeting of its Administrative Council on 23 May of that year Dr Pedro Álvarez, a veteran IAC astrophysicist and technologist, was named Director General of the company.

At around that time Spain's leading astrophysicists were gathered in Madrid, and I seized the opportunity to discuss the project and its problems with them, assuring them that the concept of a large Spanish telescope was now viable. At the general assembly of the Spanish Astronomical Society, held in Alicante in September 1994, a session was dedicated to the presentation of the 8-m telescope project (as it was then). Pedro Álvarez (Project Director), José Miguel Espinosa (responsible for scientific aspects and coordination with the scientific community), Casiana Muñoz Tuñón (in charge of site prospection), and myself (overseeing everything) attended the session. I made the following note on the session:

> An animated debate on the project started. The various speakers answered questions put to them by those assembled. There was clear uneasiness concerning the magnitude of the project and the burden of responsibility that this would assume, although many of those present either openly or privately expressed their interest in participating and urged that channels be established for their participation.

In sum, by the end of that year the immense majority of Spanish astrophysicists had publicly declared themselves in favour of the daunting project. This acquiescence, in a country so given to turf wars, was of great importance, for this Big Science project on which we were embarking had to carry, right from the start, the stamp of a national endeavour.

Being aware of our lack of experience, we sent our conceptual project for an 8-m monolithic mirror telescope to the few persons that had demonstrable knowledge at that time of building large telescopes and we invited them to a critical meeting at the IAC for their evaluation of our project and its viability. At the beginning of October 1994, a conference was held in Tokyo on instrumentation for large telescopes, which was attended by those already involved in our large telescope to gather information and establish contacts with those in the know. I was unable to attend, but going over the matter in my mind, as ever, I came to the conclusion that, regardless of the added difficulties, it was time to take an even more daring leap forward and consider the possibility of a segmented-mirror telescope. I had a word with the greatly experienced Professor Guido Münch (who at the time resided in Tenerife) on the matter. When he voiced his approval, I rang Pedro Álvarez and told him about my talk with Guido and asked him to sound out the experts and gather opinions on a 10-m segmented mirror.

The United States had successfully built a large telescope with a 10-m segmented mirror. Until then, the largest telescope mirrors had been of a single piece (monolithic) and care had to be taken to avoid deformations due to changes of temperature and flexion caused by the inclination of the telescope. But the technical and operational difficulties in managing the immense mass of monolithic mirrors rendered it impossible to make them wider than 8 metres. Note that, while an 8-m monolithic mirror weighs some 23 tonnes, a 10.4-m segmented mirror weighs a mere 16.5 tonnes in comparison.

We began to see many future advantages in this new type of telescope. We did not want our 8 metre to be the last of the monolithic telescopes, an instrument of the past, even though it would be more manageable and easier to build. When the telescope experts met at the IAC, we presented them, as an alternative to the 8 metre, a 10-m segmented-mirror telescope project. The finest were gathered there: Jerry Nelson, Director of the American 10-m Keck and pioneer in the building of segmented-mirror telescopes; Masanori Iye, Director the Japanese 8-m Subaru; Massimo Tarenghi, Director of ESO's four 8-m Very Large Telescope; and Matt Mountain, Director of the 8-m Gemini. A very interesting discussion ensued, and the experts finally advised us to opt for the segmented design. That meeting showed the door to large monolithic mirrors.

We started afresh with a new conceptual design and its corresponding feasibility study. It soon became clear that the cost of a 10-m segmented-mirror telescope would be of the same order as a telescope with an 8-m monolithic primary; in other words, for the same price we could achieve a 50 per cent bigger light-collecting surface and, therefore, much better angular resolution.

A team of three astronomers and four engineers from the IAC spent some time in California, where the segmented-mirror Keck Telescope, now in operation, had been built. They then went on to Hawaii to talk to the astronomers and technicians using it. They went over every minute aspect of the fabrication, fine tuning, and operation of the segmented mirror. They saw the Keck Telescope in action and could verify its high observing performance. At last, after an exhaustive study of the pros and cons of the segmented versus the monolithic model, the decision was taken to opt for a segmented-mirror telescope, optically equivalent to a 10.4-m monolithic-mirror telescope, that would be optimized to observe in the visible and infrared.

Of course, in order to continue with the conceptual design of the telescope and carry out the feasibility study we had to obtain the approval of the Governing Council of the IAC, which it granted in February 1996. The

Minister for Education of Science at the time, and therefore President of the Governing Council, was Jerónimo Saavedra, one of the great Canarian politicians of the Transition. Luckily, he was a friend from the days when we were both untenured lecturers at the University of La Laguna and was well-acquainted with the IAC.

Before this, the Government had created the Assessorial Committee for Large Scientific Facilities, whose task was to analyse our telescope project. They were extremely reticent, for neither they nor anybody else in Madrid really believed us capable of building such an instrument in Spain (our atavistic complexes surfacing once again). They ended by recommending that the Government give the project the go-ahead on three conditions: that there be international participation, that the director of the project be a foreigner with proven experience in this kind of telescope, and that the necessary funding be secured before proceeding with the project. All very logical and prudent, but it left the project dead in the water.

However, in parallel we managed to convince the Ministry of Industry and the corresponding board of industry of the Canarian Government to earmark funds to award 'trial contracts' for the technology required for the large telescope. In this way, we were able to demonstrate that our firms had the required capabilities; we assembled a significant group of Spanish companies to participate in the international bidding that we would need to arrange for fabrication if we were finally to build the telescope.

With the arrival of Esperanza Aguirre as the new Minister for Education and Culture, things took a turn for the worse. I can remember only too well when, on 13 July 1998, before an IAC Governing Council meeting, she invited me, together with Professor César Nombela (then president of CSIC) to have lunch at her private ministry dining room. It was a carefully laid trap to persuade me not to proceed with the project; they tried every trick in the book. We spent hours arguing until Aguirre brought the meeting to a close with her customary high-handedness and sardonic humour, 'You'll never get that telescope built, my lad!' I am mystified as to where she and the CSIC President had got such an idea of the presumptuousness and inconvenience of the project. Some years later Nombela himself told me who had persuaded him of the need to prevent at all costs the construction in the Canaries of that 'unnecessary and impossible monster'. But I have never engaged in the habits of gossip and back-stabbing, so I refrain from repeating the name here. In fact, that very same person, later occupying a very influential post, was also among those who convinced the Minister Garmendia (and she, President Zapatero) of the 'impossibility and political inconvenience' of defending the cause of bringing the European Extremely Large Telescope (E-ELT) to the Canaries. But that is another story that will be told in due course.

We of course pushed ahead with our efforts to build the telescope. Among other things, I dedicated my efforts to extolling the scientific, technological, and strategic virtues of the project to important people, including the Prince of Asturias (Honorary Astrophysicist of the IAC). I am told that Mrs Aguirre complained about my campaign and accused me of hounding her. Try as we might, we could not succeed in breaking the Minister's veto. But I have never taken no for an answer ('He who holds out carries the day', as Cela used to say).

In 1998 President Aznar established the Science and Technology Office (OCYT), answerable to the Presidency of the Government to centralize the greater part of R&D activities. Fernando Aldana, a lecturer at the Polytechnic University of Madrid, was appointed its head. We were fortunate to have known him previously, for he had once visited us as a senior representative of the Ministry of Education and Culture. He arrived at Tenerife, as he himself said, filled with antibodies against the IAC. During his visit he was 'converted'. He said he had been warned that the IAC and its projects were all show, led by 'a public menace, a devil complete with tail and horns'; but when he met us and could talk freely with us, when he got to know us better, he saw things as they really were. He became convinced that the IAC's large telescope would produce the best and greatest impact on R&D that Spain could hope for, and that the IAC was amply capable of seeing it through.

In the end Aldana did manage, on the quiet it seems, to get Aznar to approve the project. In October 1998, the General State Administration joined GRANTECAN, S.A., and appointed as advisers to the company were: Adana himself; Pascual Fernández, Director General of Budgetary Analysis and Scheduling of the Economy and Treasury Ministry (to which FEDER funds were directly answerable); Elisa Robles, Director General of Industry and Technology of the Ministry of Industry and Energy (she would later become Minister of the Economy and Treasury); and Tomás García-Cuenca, Director General of Higher Education and Research of the Ministry of Education and Culture. Representing the Canarian Government were three of its councillors: José Carlos Francisco of the Treasury; José Mendoza of Education and Culture; and Francisco de la Barreda of Industry and Energy. I itemize all of this to underline the weight that decisions of the Administrative Council of GRANTECAN (which in 1998 Pascual Fernández would go on to chair) carried at that moment. At last, the Gran Telescopio Canarias (GTC) figured formally in the General State Budget for 1999. Owing to the inherent inertia of the state budget towards any alteration of its structure, GRANTECAN began to accumulate treasury residues that permitted its Director to under-take the construction of the telescope and keep the first years of its operation solvent. There can be no doubt that Fernando Aldana, fully in favour of the project, put his heart and soul into helping us make it a reality. In the two years that he ran OCYT, before that entity gave way to the new Ministry of Science and Technology, GRANTECAN became economically secure, but it must be added that he faced stiff resistance. For that reason, he has a well-earned place among the key figures of the GTC. Having fallen from grace sometime later, he was overlooked in the inaugural celebrations. Thus are some repaid for their efforts. But he is not forgotten at the IAC, and we were able to get the International Astronomical Union to get an asteroid[2] named after him.

To summarize, in 1996 we began the conceptual design, published in May 1997, and followed on with the preliminary design. The detailed designs were undertaken from 1999, with funding now secured. That same year we signed

Fig. 21.1 The author explains the workings of the future GTC to HRH Prince Felipe in the presence of the President of the Local Canarian Government. (Credit: IAC)

[2] Minor Planet 44103 (Aldana).

contracts for the principal subsystems of the telescope. On 2 June 2000 we ceremoniously placed the first stone in the foundations of the pillars of the telescope. The ceremony was presided over by Felipe de Borbón, Prince of Asturias, who had always given his support to the project. In 2001 the principal astrophysical institutions of Mexico and the University of Florida, both with a 5 per cent participation, were formally incorporated into the project, thus finally giving the project an international character.

In 2002 we began an initiative that aimed to stimulate scientific teams from the GTC partner countries immediately to weigh and plan the science they wanted to do with this telescope as soon as it entered into operation. The first *Science with the GTC* conference was held in Granada that same year. In 2004 we would celebrate the second one in Mexico City, and the third two years later in Florida. We are now planning the sixth in the series. These meetings are a useful tool that encourages the sharing of core principles, contacts, and synergies.

To ensure that the construction would be carried out rigorously we formed a small committee with the best experts in the world in this specific field. The committee was chaired by Jerry Nelson, who had built the only two then existing segmented-mirror telescopes, the 10-m Kecks in Hawaii. We contracted the team to make regular detailed inspection visits of the project and prepare, after each visit, a written report with criticisms and mandates. Their exigent demands kept the project on a sure footing.

Needless to say, the years of construction work were plagued with problems and difficulties, as was only to be expected in a project of such dimensions, to which were added further problems arising from undertaking such a project in Spain. But solutions were found, one by one, to all these hiccups. Some of these obstacles were extremely serious, such as having to terminate a contract with a joint venture involving a group of prestigious Spanish companies for the building of the telescope dome; the work had finally to be carried out by GRANTECAN's own technical team. Another non-trivial challenge was resisting blackmail by a powerful French firm that had been contracted to supply the secondary mirror. We pressured them to meet their commitments in terms of both the cost and the form of the secondary.

We all learned a great deal, especially the Spanish firms involved, which increased their capacity and competitiveness. Looking back on it all, it is clear that it is only with faith, courage, and decision, in addition to having good people and luck, that great things can be achieved. The staff forming the design and construction teams, most of them young, were accommodated in a prefabricated wooden hut on IAC premises in La Laguna; they did an excellent job, as is borne out by the excellent performance of the GTC. This good

Fig. 21.2 HRH Prince Felipe laying the first stone of the GTC, introducing a time capsule with historical documents of the year 2000. (Credit: IAC)

work is now being continued in La Palma with the observing teams currently exploiting the full scientific potential of the telescope in the most precarious economic circumstances. Every time I think of all this I want to go and congratulate these people all over again.

What fine people we have! Thanks to them, Spain has been able to build this marvellous instrument and enter the world of Big Science through the front door. For several years to come the GTC will remain the biggest and most advanced optical–infrared telescope in the world.

On the night of 11 July 2007, the telescope officially saw 'first light', as a telescope's baptism by light is termed; that is, when it was launched into the sea of photons without sinking—in other words, when it is rendered capable of seeing the stars. First light occurs when the beam of photons collected by the telescope's objective is successfully concentrated in the focal plane to form an image of a celestial body. It was a decisive and exciting moment.

When he laid the first stone, the Prince of Asturias—himself a keen aficionado of astronomy—said that he wanted to be present at this moment. And there he was, accompanied by the President of the Canarian Government, the Minister of Education and Science, and senior dignitaries of both administrations, the Cabildo of La Palma, and the municipality of Garafía, our partners from the University of Florida, the Institute of Astronomy of the National Autonomous University of Mexico, and the National Institute of Astrophysics, Optics, and Electronics of Mexico Also present were the Management Committee of the IAC and other astrophysicists, among them Brian May, the famous Queen guitarist.

After a light supper and a few talks, we went up to the telescope building and examined the large instrument in detail. At around midnight, with all of us in the control room, manoeuvres were initiated to point the telescope at the Pole Star. Next, the separate images of the star formed by each of the hexagonal segments that make up the large primary of the GTC were stacked on top of each other. It was most exciting to watch, as each segment was adjusted, how the initial multiple Pole Stars were blended into a single image. When this was achieved spontaneous applause broke out. Amidst emotional laughter and congratulations, the moment was toasted with Canarian sparkling wine. Before retiring for the night, the guests were served with soup and hot drinks at the Residence. A night to remember, forming one of the best memories to salve the wounds from so many battles.

Astronomical observing commenced in March 2009 with the first available instrument, OSIRIS (an intermediate-resolution optical spectrograph with imaging capabilities). With the telescope now in operation, its inauguration

Fig. 21.3 Inauguration of the GTC at Roque de los Muchachos Observatory in 2009. (Credit: IAC)

was scheduled for a date acceptable to the Royal Palace. The date finally decided on was 24 July 2009.

The inaugural ceremonies were presided over by the King and Queen of Spain, and commenced with a detailed visit to the GTC, with Dr Álvarez (its Director) and myself acting as guides. Inside the dome, they greeted the technicians who had built the telescope, signed the Visitors' Book, and received as a memento a beautiful image of the galaxy M 51, taken by the GTC. Outside, in the presence of more than five hundred guests, a commemorative monolith was unveiled. The dignitaries were seated on a dais along with persons who were going to give discourses: myself as founder and Director of the IAC; Professor Joseph Glover, Chancellor of the University of Florida; Professor José Narro, Chancellor of the National Autonomous University of Mexico; Mrs Cristina Garmendia, Minister of Science and Innovation; His Excellency Paulino Rivero, President of the Canarian Government, and His Majesty King Juan Carlos I.

The ceremonies at Roque de los Muchachos Observatory ended with two group photographs: one of the VIPs and the other of the invited public.

The ceremonies were televised live all over the world. In Spain alone the broadcasts reached 8.4 per cent of national viewers, and the ceremonies were covered by 30 news media.

A little later, in Santa Cruz de la Palma, the King and Queen inaugurated the exhibition *Cosmovisiones* at Salazar Palace, which demonstrated the evolution of our conceptions of the Universe through time.

The most scientific part of the inaugural ceremonies was the seminar 'Science with large telescopes', in which took part the directors of the largest terrestrial and space telescopes.

It is worth noting that the GTC did not overrun its budget—a rare occurrence with Big Science projects. Large public works often double or triple their original budget. The construction of the GTC, including its first two instruments, cost around 105,000,000 euros. When adjustments are made for the rate of inflation, this amount amounts to 10 per cent below the initial budget. Neither had we deliberately inflated the true cost, as is clearly borne out by the refusal of some European countries to participate in the GTC project on the grounds that it would be totally impossible to build the telescope with such a low budget. Germany in particular gave this very reason for not entering into the project and decided instead to spend its money by allying with the Americans to build the Large Binocular Telescope (LBT) in Arizona (a project that suffered many delays and cost double the initial budget estimate). During the LBT's inauguration, with the telescope still not finished, I could not refrain from teasing my German colleagues on the irony of it all.

One thing we were unable to do was meet all the set deadlines. There were some four years of delays caused by the proactive preliminary estimates that were quite honestly made in a spirit of optimism and eagerness to get the GTC up and running as quickly as possible. Even so, counting from 'first stone' in 2000 to first light in 2007 (these are the usual milestones in making comparisons for this kind of project), only seven years were invested in the construction of the GTC, about the same length of time needed for the Keck telescopes in Hawaii and the VLT in Chile.

Then came the lean years of the economic crisis, reversals of political polarity, and party discrepancies between Madrid and the Canaries. Before the finishing touches could be given to the telescope, the security of multi-year budgetary commitments agreed between the central and autonomous administrations came to an end. To the technical difficulties of finishing such a complex undertaking were added economic ones. This nightmare has had to be endured with the greatest difficulty and continues to hamper and bog down the proper running of the GTC, taking the edge off the telescope's scientific competitiveness.

From 2010, the Gran Telescopio Canarias ceased to figure in the General State Budget. In subsequent years its running, maintenance, and updating costs have been covered by funds corresponding to the finalization of the

Fig. 21.4   The Gran Telescopio Canarias. (Credit: Pablo Bonet/IAC)

construction of the telescope. Its economic situation is becoming more and more insecure.

To run a large scientific facility of such distinction it is necessary to invest around 10 per cent of the cost of its construction in maintenance, but the GTC is having to get by on barely 4 per cent. That is typical of how things are in Spain: backbreaking effort goes into building something, and then there is no money to keep it working.

Consider one example more of the parsimony of governments in scientific matters, whatever their political hue. In its poorly handled negotiations concerning Spain's entry into ESO, the government, mindful of keeping costs down, agreed a payment in kind of 40 nights per year on the GTC over a period of five years (valued at 15,000,000 euros). The GTC never received compensation for this, in spite of the transaction being an outright sale of telescope time that should therefore be considered an asset of the GTC. Had the GTC been repaid this debt, it would not be in the distressing economic situation in which it now finds itself in its struggle to guarantee normal service.

Another irritant is the different treatment in scientific and industrial matters meted out to the Canaries in comparison with Catalonia and the Basque Country. When Spain lost the International Thermonuclear Experimental Reactor, a large international project attempting to show that producing energy through nuclear fusion is viable, Catalonia received generous compensatory funds to build ALBA, the Catalan synchrotron. Similarly, when Bilbao lost out on the European Source of Neutrons by Spallation project, the Basque Country received generous funding for its research infrastructures. In the case of the Canaries, however, when Spain lost the E-ELT, the Canaries did not receive a whit of compensation, and the GTC was denied the necessary funds for its operational costs. A truly incredible state of affairs!

In spite of all these setbacks, the strategic technological objectives laid down with the GTC project have all been convincingly achieved, and its scientific objectives are being fulfilled. More than two hundred articles have been published in international peer-reviewed journals, and the rate of growth of publications is similar to that achieved by comparable telescopes. It is amazing how hitherto unseen details are being detected with our telescope. The birth and evolution of distant stars and galaxies are being witnessed, and chemical elements produced shortly after the Big Bang are being found in unexpected places. The GTC's fine resolution of detail is resulting in highly significant advances in our knowledge of the extragalactic and nearby Universe—in the behaviour of black holes, for example. Also noteworthy is the detection of potassium and sodium in exoplanetary atmospheres. To give an idea of how extraordinary this detection is in an exoplanet so close to its parent star, imagine being able to distinguish from Neptune a lighted match in New York with the city in flames. The detail seen on asteroids and comets will be help us better understand the origin of the Solar System and even the presence of water on earth. What else might be discovered if there were the funds for the full exploitation of the GTC?

This miracle is the fruit of the hard work of the many people who give their daily and nightly support to the GTC, especially the sleepless nights of its two directors: first Dr Pedro Álvarez and now Dr Romano Corradi. Neither should the long battle permanently waged by the IAC to secure funds be forgotten; these efforts have made it possible for the Gran Telescopio Canarias to continue observing: without those efforts there would have been no other option than to close it down.

It is also just to recognize and extol the patriotism and generosity of the Canary Islanders, particularly in these politically centrifugal times, when Spain's autonomous regions have developed a selfish tendency to contemplate their own navels. Ever since the creation of GRANTECAN, S.A., the Canarian Government has provided significant funding to this project, without demanding observing time in return or other privileges for their local compatriots. And the same may be said of the IAC.

It is only fair to end this chapter by making a special mention of three people who have been key to the GTC project. But first let me underline in red that to build and put into operation this large telescope is in itself an act of prowess, but to achieve such a thing is Spain is a feat of epic proportions. The three persons I am going to mention are quite different from one another but curiously they have a basic training in common: they are all physics graduates. Because they graduated before astrophysics was taught in Spanish universities, they followed different career paths. Ramón Ascanio Togores took up technology in a private company, José Miguel Rodríguez Espinosa became an astrophysicist in the United States, and Pedro Álvarez Martín trained as an astrophysicist at the IAC, his career soon following a path dedicated to instrumentation when he became head of the IAC's Instrumentation Division.

From the commencement of its construction to the commissioning of the GTC Dr Rodríguez Espinosa, with his good nature, professionalism, rigour, and patience, has brought about the effective participation of the Spanish astrophysical community, and has seen to it that the telescope has not fallen short of the scientific performance expected of it. Next, one has to focus the spotlight on a person who, behind the scenes, has always continued to guarantee that the administrative and managerial machinery ran smoothly in truly difficult conditions in such a country as ours. I refer to Ramón Ascanio:

efficient, discreet, and honest through and through. The third key person is Dr Álvarez, Director of the project, and passionately devoted to it, long before it became a reality. He underwent an unlikely conversion from research astrophysicist and technologist to company director, and not just any company but a public one specializing in high technology in Spain, involving participation by the State and the Canarian Community, and—as if all that were not enough—one dedicated to undertaking a project that was considered to be impossible. There are no words to express it. Those who have not endured the combined administrative tribulations of State and Autonomous Community, the extreme exigencies of astrophysicists for their instruments, and the voracious jungle of supply companies can never have any idea of the immensity of the task. He is now rightfully recognized as one of the most distinguished makers of large telescopes. Today you will find him in retirement, with not a care in the world, tending his garden in Las Breñas or taking an afternoon stroll anywhere in La Palma, wearing his Canarian farmer's cap and always accompanied by his wife, the wonderful Manoli.

22

The European Extremely Large Telescope: Dealings with ESO (1961–Present)

The European Southern Observatory (ESO) is the largest European intergovernmental organization in astronomy. Its telescopes are located in the Atacama Desert, La Silla, and Paranal in Chile, and its central headquarters are in Garching, just outside Munich. ESO was formally established towards the end of the fifties, when astronomy was not in a very developed state in Spain. In any case, the question of basic research did not arouse any particular interest, so Spain did not participate in ESO.

At present the participation of ESO in the Atacama Large Millimeter/submillimeter Array (ALMA), currently the largest ground-based astronomical instrument in the world, is particularly important. Located in Llano de Chajnantor, 5000 m above sea level, ALMA is a complex international astronomical facility and is the result of a collaboration between ESO, the US National Science Foundation (NSF), and the National Institutes of Natural Sciences of Japan (NINS) in cooperation with the Republic of Chile.

The present challenge to ESO is to build the European Extremely Large Telescope (E-ELT), which will have a 39-m primary mirror.

In what follows, I shall describe what it cost for Spain to form part of ESO.

From the time that Spanish astrophysics began to develop a collective consciousness of its own existence, we pushed for Spain to enter this European organization. We may date this collective awareness from 1975, when we who were dedicated to astronomy in Spain in one form or another gathered in the Canaries at a meeting entitled *First National Assembly of Astronomy and Astrophysics*. By then, there were already a number of doctors in astrophysics and the first chair in astrophysics had been established at a Spanish university; there were two astrophysics institutes, one in Andalucía and the other in the

© The Author(s), under exclusive license to Springer Nature Switzerland AG 2021
F. Sánchez, *The Rise of Astrophysics in Modern Spain*, Astronomers' Universe,
https://doi.org/10.1007/978-3-030-66426-8_22

Canaries, and teams of astrophysicists began to sprout up in Spanish universities.

Nevertheless, from 1961 there had been some sort of relation with ESO. I myself had been awarded a grant by the International Astronomical Union to go to France and Belgium at the beginning of 1962 to learn the techniques that ESO had been employing in South Africa and Chile in its site-testing campaigns. I learned these techniques, including the methods of data reduction and analysis, and presentation of data from Dr Dommanget. It was a double grant, and this Belgian astronomer came to the Canaries to complete the collaboration. As a consequence of all this, the same methods were employed in our campaigns, all, of course, dependent on the slender means available to us. This similarity of methodology enabled us to compare similar parameters, with the result that we soon became convinced that the quality of the skies above the Canarian summits was similar to that of the best sites prospected by ESO.

Contacts and collaborations between Spanish astrophysicists and ESO became more frequent as astrophysics began to flourish in Spain. However, we had to wait until no less than 2006 before our country joined this European scientific organization. Apart from the lack of interest among our politicians for basic science, the reason for such a long delay may be attributed to the extremely high price tag carried by the powerful and brilliant international organization that is ESO. It has potent means of observing and an extensive, complex, and costly infrastructure spread over two continents. For that reason, entering as a new member of, and remaining in, the organization costs a great deal of money. That is why successive Spanish governments were always postponing a decision on entry. Only in the twenty-first century, with a solid, successful, and well-established astrophysics in Spain, could we wear down the reticence of the administration. I remember the argument that definitively swayed the Government being the Gran Telescopio Canarias, then at an advanced stage in its construction. There were two reasons for this change of mind: first, part of the high cost of Spain's entry could be defrayed by granting observing time on the GTC; second, thanks to the GTC, the Spanish companies that had been participating in its construction were now prepared to compete in the call for the rich work tenders issued by ESO. This was later clearly demonstrated. There was a third, very powerful, reason: to strengthen Spain's candidacy to host the E-ELT. That is why the entry agreement explicitly states that attempts would be made to ensure that the E-ELT be built in the Canaries. I draw attention to this now in anticipation of what eventually happened in this regard.

In the first decade of this century, with its large optical–infrared telescope on the point of entering into service, Spain had become an astronomical power, and ESO was keen that our country should not remain outside the fold. Without Spain ESO would be incomplete. Apart from reasons to do with image and other considerations, there were economic factors too: in terms of ESO's budget Spain was very important because its contributions would be based on its gross domestic product. There were also important complementary reasons regarding ESO's plans for the extremely large telescope, such as knowing at first hand the technological advances incorporated into the GTC. Moreover, many European astronomers lacked access to modern telescopes in the northern hemisphere and were keen to access those in the Canaries. With all these aces in our hand the negotiations should have been to our advantage. It might seem incredible that the Spanish negotiators even managed to lose us the E-ELT, but it was by order of the Ministry of Science and Innovation that those with the most clearly demonstrated experience in these kinds of international negotiations were expressly forbidden to join the negotiating team, on the fatuous grounds that they were from the IAC and could not therefore be deemed independent.

I need to return to the last century in order to explain something that turned out to be quite important in our relations with ESO. Since the signing in May 1979 of the treaty that opened the IAC Observatories to the international community, the summits of the Canaries have served as a natural European enclave for ground-based observing in the northern hemisphere. These Observatories, as far back as the eighties, had sound legal bases and the means to be offered as an ideal multipurpose observing platform for the most advanced telescopes. Furthermore, astronomers from many countries, after many years of working together in the Canaries, had built very stable synergies and habits of pan-European cooperation. We began to speak of and promote a European Northern Observatory (ENO), a name that did not at all sit well with the people from ESO.

ENO was conceived as a 'European facility' of the European Space Area (ERA), a flexible association model with 'variable geometry', an idea that the European Union was promoting at the time, as the Commissioner himself, Mr Busquin, explained to us on his visit to the Canaries. We even managed to have explicit references to ENO made in the VI Framework Programme of the European Commission. This raised yet more hackles at ESO.

ENO had historical antecedents. In 1967, Hermann Brück, Astronomer Royal for Scotland, had already proposed the construction of a 'Northern Hemisphere Observatory' in the Canaries, based around a British 150-inch (3.81-m) telescope. Exactly two years later, a committee chaired by Fred

Hoyle was given the task of studying the viability of such a project and gave its approval. In the middle of 1971, the British began its planning and carried out further astronomical site testing at Roque de los Muchachos Observatory. They were consequently the first signatories of the Treaty of Cooperation in Astrophysics in 1979; since then, they have set up their best telescopes in the northern hemisphere at this Observatory.

In our innocence we believed that such a pro-European idea as ENO would enjoy the solid support of all the organizations with telescopes in the Canaries; we had forgotten that each organization has its own strong institutional interests, wih even personal motivations involved in the case of ESO. Our beautiful dream was finally to have ENO recognized as such by everybody with a view to then negotiating the creation of an entity to be known as the European Observatories (EO), effectively the joining of ESO and ENO. The EO would be a multinational organization encompassing all European ground-based observing facilities in the same way that ESA embraced all European space activity. But we had not discussed and agreed this strategy beforehand with ESO. Bad mistake.

As the Canarian Observatories continued to prosper, we began to aware of ESO petulantly heaping opprobrium on us, perhaps a normal response by those secure in their power when disturbed at the emergence of newcomers in their domain.

The IAC and its Observatories (the basis of ENO), with their efficient, low cost structure certainly stood in stark contrast to the high cost of ESO, an uncomfortable feature that the latter might feel to be its Achilles' heel. And so it has continued, as recently happened with the selection of the site for CTA North, an extensive international battery of Cherenkov telescopes. Our experienced site-testing team had yet again to fight hard to expose the manipulations contained in the comparative site-testing studies for this array. The matter ended with ESO, rather than Argentina, being given CTA South, CTA North going to Roque de los Muchachos Observatory (with the inauguration of the four Japanese Large-Sized Telescopes the decision is now irreversible).

Matters reach boiling point with the arrival of the Franco–Argentinian Catherine Cesarsky as Director General of ESO, and the ill feeling continued during the tenure of Tim de Zeeuw—another cause that led to the manoeuvres preventing the E-ELT from being built at Roque de los Muchachos Observatory. We hope that these attitudes, which are to nobody's advantage, will disappear now that the new Director General of ESO, Xavier Barcons, is a Spaniard. Let us hope so.

I mention these confrontations at the risk of boring those readers with no interest in them, but historians of science have insisted that I do so because these are important matters that will serve to understand better the evolution of European astronomy in this century, so I shall try to be concise and not offend anybody.

Things were different when Riccardo Giacconi was Director General of ESO (from 1993 to 1999), so much so that the EO was indeed about to be initiated, ESO thereby establishing itself in the northern hemisphere while at the same time settling Spain's entrance into the organization. The human and scientific qualities of this future Nobel laureate transcended all meanness. He knew ESO inside out and was the first to tell me, with some bitterness, things about the organization that needed to be improved, although there was no way of achieving this. One of these problems was that a new management committee was set up when there was a change of general director, the new incumbent surrounding himself with people of confidence but not having the powers to make a clean sweep of older members of the committee put there by his predecessors. Thus was created the burden of a cumulative committee that that has been both costly and conflictive.

In the course of a working visit by Giacconi to the IAC, with which he was well acquainted, he had time to appreciate important nuances during lengthy talks and friendly working meals in picturesque spots in the Canaries. Together, we made a detailed analysis of future solutions to provide a better service for European astronomy, and we arrived at practical and workable solutions. We could see that the best thing for everybody would be for ESO to begin to establish itself also in the north, at the Canarian Observatories, while Spain would at the same time gain entry into ESO at a cost that would not be disproportionate. This would be right road towards endowing Europe with optimal observatories in both hemispheres at a reasonable cost. We understood this solution to be viable on the grounds that ESO would participate in the large telescope that the IAC wanted to build at Roque de los Muchachos Observatory, the new telescope serving as Spain's entry fee into ESO, very similar to what Britain had done in La Palma. A high-level joint Spanish–ESO negotiation committee was formed. Just as all was going well and to the satisfaction of all, Giacconi left the general directorship of ESO, while in Spain the Office of Science and Technology (OCYT), which was directly answerable to the Presidency of the Government, was wound up. Everything came tumbling down catastrophically.

We were unlucky. The Spanish–ESO joint negotiating committee met in Madrid, and afterwards Catherine Cesarsky, the new Director General of ESO, met with Fernando Aldana, then Director of OCYT (Chap. 21). Aldana, who was sure he would in a few days be given the Ministry of Science and Technology, was euphoric and gave vent to his peculiar brand of sledgehammer humour. Cesarsky misconstrued his clumsy humour when he said to her, 'How much are the observatories in Chile worth? I'll buy them from you so we can end the negotiations once and for all.' Up to that point the negotiations had indeed been about to end on a positive note. Aldana placed his car at Cesarsky's disposal, and I accompanied her to the airport. She was most put out, and I could do nothing to convince her that it had all been nothing more than a misguided joke. Twisted things stay twisted. Cesarsky felt affronted and was in no mood to appreciate subtleties or easily forgive.

It also needs to be borne in mind that, since 1993, I had never been popular at ESO. I was invited to give a talk at the Catholic University of Santiago de Chile, and I took the opportunity to make a detailed tour of the European and American observatories in that country. I detected in them an underlying scientific and technological colonialism that I did not like. I committed the cardinal sin of telling the Chileans how we had done things in the Canaries to keep our observatories Spanish. It needed to be understood, I told them, that a sky of high astronomical quality was a 'natural public resource', which was not something that should be sold along with the lands that lay beneath it. I explained to them how on that basis we had negotiated advantageous collaboration agreements that were decisive in helping us raise the level of Spanish astronomy. I seem to have opened the eyes of the new Chilean authorities, who decided to renegotiate the agreements with the foreign astronomical institutions. Pinochet had been thrown out of power to make way for democracy. At a later date the Chilean Vice-minister of Universities, Science, and Technology visited the IAC and its Observatories. We spoke openly on a wide range of issues. He forced a renegotiation for a new agreement with ESO, and I keep as a memento the telegram he sent thanking me, after he had successfully negotiated a better return for Chile. It seems that some in ESO have not forgotten that.

Astrophysicists have an insatiable thirst for photons. That is why, as explained in the previous chapter, they need telescopes with wide-aperture mirrors to reach the deepest confines of the Universe and view celestial objects in greater detail. Hence, in nineties there were already projects for giant telescopes on the drawing board. The question was to determine up to what size such telescopes would be technically viable. A further challenge was to find sites with the sky quality demanded by such refined instruments.

The technology and experience acquired in the building of telescopes with segmented primaries, such as the Gran Telescopio Canarias, had opened the path to the design of telescopes with extremely large primary mirrors: the so-called extremely large telescopes (ELTs). Two projects arose in Europe: a fantastical one with a 100-m primary, the OverWhelmingly Large telescope

(OWL) promoted by ESO and destined for Chile, and another with a more reasonably sized 50-m primary, the EURO-50, promoted by Sweden and Spain, to be built at Roque de los Muchachos Observatory. The latter project was to prove a further bone of contention with ESO.

Carlos Westendorp, then chairman of the Industry, External Trade, Research, and Energy (ITRE) Committee of the European Parliament (he had formerly been Minister for Foreign Affairs in Felipe González's government) took the projects for European giant telescopes to be debated at the European Parliament. On 27 May 2002, presentations were made by Catherine Cesarsky, Director General of ESO, flanked by a cohort of assessors, and myself as Director of the IAC, accompanied only by my secretary, who was at the time on a service commission in the European Commission. ESO's Director General spoke of the wonders of her powerful organization and of the future glories of OWL, once set up in Chile. In my presentation, I showed how only the EURO-50 was technically feasible at that time; by building it in La Palma, I insisted, where there already existed an extremely efficient multinational infrastructure and organization, effectively the European Northern Observatory, a great deal of time and money could be saved. Moreover, I said, being operational very quickly, it would give Europe the edge over a similar American project. The two postures were debated and everything was left pending further consideration, as is habitual in the

Fig. 22.1 Artist's impression of the 100-m OWL (OverWhelmingly Large telescope) proposed by ESO Telescope Systems Division. (Credit: ESO)

European Union. But public mention of ENO in the same breath as ESO at European level only increased the unease and resentment of ESO.

From that point onwards, ESO began systematically and shamelessly to anathematize the IAC Observatories, as I now explain, basing my account on fully archived documentation on the matter. In the view of those representing the European Southern Observatory, the hour had come to remove from their midst the threat that ENO posed to their prestige and hegemonic designs. The effect was felt immediately in numerous ways. There were soon bitter and angry exchanges at CCI meetings, whose membership included persons serving on the governing organs of ESO. It did not take long for them to ensure that ENO should end up dissolving through inanition. In the absence of divine intervention in this battle between David and Goliath, it was David who stood to lose everything.

But there was more to come. A technical working group was set up with European funding to examine which of the two technical proposals, EURO-50 or OWL, would be the more suitable for the European extremely large telescope. Needless to say, ESO managed to stifle the EURO-50 in favour of its OWL, however much the Spanish and Swedish members of the group demonstrated in tough technical meetings the clear advantages offered by the EURO-50. What happened next? First, ESO, in the face of the multiple construction problems with OWL that rendered it unviable with a 100-m mirror, were forced to reduce its diameter to 60 metres. But the project still remained impracticable, so they made a further reduction to forty-odd metres. That was still not doable, with the result that today the diameter has shrunken to a 'mere' 39 metres. This version is now the E-ELT, whose construction has begun in Chile.

After the EURO-50 had been wiped from the slate, we focused all our efforts on getting the E-ELT sent to the ORM, knowing all the while that ESO wanted it in Chile. As is usual in the IAC, when we face a major challenge we set up a think tank. We have always believed in brainstorming. The group consisted of our best researchers and was set the task of analysing the subject in depth, devising a workable strategy, monitoring the evolution of events, and proposing practicable solutions to be tried at all times. We knew we were up against a giant, so we had to have courage and determination. We were encouraged by knowing that this would not be the first time we had achieved impossible goals. We were also sure in our knowledge that we wanted the best for the Canaries, for Spain, and for Europe—and for astrophysics too. All or nothing! You cannot enter the arena with the spirit of a loser.

We were quite aware that the final emplacement of the E-ELT was going to be decided on political and economic grounds, and also on personal

Fig. 22.2 Artist's impression of the EURO-50 telescope at Roque de los Muchachos Observatory. (Credit: Lund Observatory/Science Photo Library)

considerations, with the astronomical quality of the site taking a back seat (although it would be broadcast as the primary factor under consideration). We endeavoured to cover all these aspects through specific actions. We had to promote our best selling points efficaciously and sway those minds that we could. It was also vital to act quickly. It was not going to be enough to argue that the IAC Observatories, particularly the ORM, were the best astronomically characterized sites in the world, with more than thirty years of rigorous site-testing data published, and more than forty years of permanent astronomical observing on the telescopes of more than sixty scientific institutions from nineteen countries. In spite of all these points in our favour, we decided to make further specific observations with the latest techniques and instruments. We also needed representation in all forums and groups dedicated to these topics. And we had to achieve all this with a very small staff. It is incredible that we were able to do almost all of it. Special praise is owed to Casiana Muñoz Tuñón, Toñi Varela, and their site-testing team.

Since one of ESO's hobby horses was the claim that the southern hemisphere was more important for ground-based astronomical observing, we mooted the organization of a meeting of leading astrophysicists under the title *Science with the E-ELT from the Northern Hemisphere*. The meeting was held in Madrid on 16 and 17 April 2009, with personnel from ESO also attending. A number of science cases on celestial bodies of great significance were analysed, and it was concluded that there exist unique astronomical objects in both hemispheres, and that top-level science (which must, beyond doubt, be the prime objective of the E-ELT) could be carried out regardless of the hemisphere chosen.

We had a further intention of motivating Spanish astrophysicists, who were well represented at the meeting. They gave their clear and strong support for sending the E-ELT to the Canaries, to the ORM.

The proceedings of the meeting were published in English. This book is still of great interest, as demonstrated by references made to it by the American Thirty Meter Telescope (TMT) in reaffirming its emplacement in the northern hemisphere, in either Hawaii or the Canaries.

When we began promoting the Gran Telescopio Canarias project, it had been necessary to persuade the Spanish astronomical community to regard the GTC as its own; we again appealed to that same community on every occasion to secure the E-ELT for Spain. It almost happened. I remember when, in 2009, representatives of every Spanish astronomy and astrophysics research centre backed the demand: 'This community has declared clearly and enthusiastically its keen scientific interest in the project'. Also in that same year the National Commission for Astronomy, whose purpose, according to the Official State Bulletin, is 'the promotion and coordination of national astronomical programmes, [and] assessment on behalf of the General State Administration in the matter of Astronomy and Astrophysics', recommended unanimously that everything necessary be done to ensure that the E-ELT be brought to Spain.

The Prince of Asturias (now King Felipe VI) presided over the inauguration in Spain, on 27 January 2009, of the International Year of Astronomy of the United Nations. In his discourse he was also clear:

> In the end, science is the ultimate *raison d'être* for these prodigious machines that technology is placing in the hands of researchers, and through them—as with the GTC—we know that our companies have achieved advances that are making them more competitive. We know that they are likewise involved in the preliminary developments of the super telescope of the twenty-first century, the E-ELT (European Extremely Large Telescope) that is now under way. We hope that, when the time comes, sufficient astronomical, logistic, cultural, and economic reasons will converge to make a reality our aspiration that it finally find

a home in the Canaries, in Europe's natural observatory in the northern hemisphere.

A detailed socio–economic study of the repercussions the emplacement of the E-ELT in La Palma would have on the local environment was commissioned from specialized firms. Their conclusions were dramatic. At a most conservative estimate, the returns could mean an added value of some 720,000,000 euros for the Canaries and more than 1,100,000,000 euros for Spain generally. Regarding the impact on employment, 750 full-time jobs would be generated annually in Spain during the eight years of the construction phase, of which 51 per cent would be located in the Canaries. And during the minimum of 30 years that it was estimated that the super telescope would be in operation some 700 jobs would be created annually in Spain, 37 per cent of which would be in the Canaries. These figures would be notably improved if the negotiations with ESO went well.

In economic terms, the overall costs for ESO of the building and operation of the E-ELT in La Palma, rather than on the inhospitable peaks of the Chilean desert would be 30–40 per cent lower. The logical advantages of remaining in Europe, in a fully equipped observatory with developed European communications and infrastructure greatly favoured Spain over Chile. Apparently, it was not considered opportune to lay that particular ace on the table.

Fig. 22.3   Artist's impression of the E-ELT (European Extremely Large Telescope) now being built at Cerro Armazones in the Atacama Desert (Chile). (Credit: Swinburne Astronomy Products/ESO)

The report was submitted to the Ministry of Science, Technology, and Innovation, and the Ministry of Economy and Treasury, and also to the Canarian Government. The political parties were also informed. The question of the E-ELT was raised in the Canarian and national parliaments on repeated occasions. Yet the Ministry of Science, Technology, and Innovation always voiced its opposition. Looked at now through the perspective of time and taking into account all that had happened then, this opposition is of particular significance.

Nevertheless, 300,000,000 euros were placed on the negotiating table, the amount that ESO said was still needed to build the telescope. A non-legislative proposal prospered and was supported by all the political parties, the PSOE included, thus obliging the Spanish government, during the Mr Zapatero's presidency of the European Union, to back actions to get the super telescope built in the Canaries. The E-ELT was then the only European large scientific infrastructure of ESFRI (European Strategy Forum on Research Infrastructures) with any possibility of being located in Spain. There is also documentary evidence that the European Parliament broke a lance in favour of installing the E-ELT in the Canaries.

In spite of everything the E-ELT went to Chile. On 4 March 2010 the Council of ESO decided it would be located on Cerro Armazones, a mountain for which an access road would need to be built and whose summit would need to be levelled. But ESO has never wanted to publish the scientific data on which their decision had been based. The astrophysicist representing Spain on the Council, Xavier Barcons, when presented with our demand that Spain should force the comparative data to be published before making the Council's decision public, said that he did not do this 'in order to prevent the fame of the astronomical quality of the Canarian sky being torpedoed'. A strange reply, even stranger in the light of what is known today, thanks to the American TMT team. ESO finally made its decision public on 20 April.

If historians should someday decide to investigate the reason why Spain, with so many trump cards in its hand, came to lose the E-ELT, they will find abundant material for their research. There are the minutes of meetings of the Spanish and European parliaments, together with those of the Council of ESO, and the official registers of the coming and goings of Spanish ministries and the IAC, thanks to the detailed coverage of the topic by the news media. Here, I have simply given a brief account of a number of significant episodes that speak for themselves.

For a number of years there was a working group, funded by the European VI Framework Programme, dedicated to 'E-ELT Site Characterization', whose mission was to make precise comparative measurements worldwide, particularly in the Canaries and Chile, with the aim of finding the best site for this telescope. Throughout this time-consuming process Spanish experts had to waste a great deal of time countering, with hard data, disparaging rumours being peddled by experts in site-testing from ESO concerning the sky above the Canarian summits. Casiana Muñoz Tuñón, one of the most knowledgeable persons at the time on matters relating to astronomical site characterization, was well informed of the tricks being used repeatedly. She had to devote much time and effort to dismantling the false rumours with measurements taken on site with modern instruments, as well as writing serious scientific articles published in international journals dedicated to the subject. Her work is there for all who care to look.

Professor Jean Vernin, chairman of the E-ELT Site Characterization Working Group, criticized what he saw as the scandalous choice of Armazones in an email sent to, among others, Tim de Zeeuw, the then Director General of ESO. He stated that Cerro Armazones was the one site not measured by this international team, created precisely to certify which site was the best for the emplacement of Europe's super telescope, and that it had in fact been previously discarded by the Americans for the TMT. He emphasized in his email that Armazones was 'the only site we have not probed, and the which has been abandoned by the TMT group who preferred Mauna Kea'. He ended by complaining that ESO was also making confused references in the press to results from a certain site-testing team that was certainly not the one led by him.

Should there be any remaining doubt that ESO did not reject Roque de los Muchachos Observatory on scientific or economic grounds, there is the telling evidence that the United States, in the light of problems in Hawaii, has now selected Roque de los Muchachos Observatory as an alternative site for its Thirty Meter Telescope. Their decision was made public on 31 October 2016 and is still in force, even though the Supreme Court of Hawaii has granted permission for the telescope's construction at Mauna Kea. The TMT leaders have said on numerous occasions that the ORM was chosen for sound scientific reasons, based on comparative measurements of sky quality. They also cite economic, operational, and security reasons. It is difficult to overlook that Chile is an extremely seismic zone.

It is public knowledge that, given various problems with ecologists and native Hawaiian sensibilities, it was decided to seek an alternative site for the TMT. After a couple of years of intense comparative studies of San Pedro Mártir (Mexico), Ali (China), Hanle (India), Cerro Vicuña, Cerro Honar, and Cerro Armazones (Chile), and Roque de los Muchachos (Spain), the Americans opted for the ORM. The *NOAO Newsletter* for March 2017 favourably reviewed the ORM's excellent conditions for astronomical observing, particularly its seeing (atmospheric stability), an essential condition for adaptive optics, which must be installed in giant telescopes such as the TMT and the E-ELT. The article also confirms that Saharan dust is not really a problem. In fact, the dust extinction statistics for Mauna Kea Observatory (MKO) and the ORM, and telescope optics are affected by dust more or less equally at both sites; however, the article states, there is more likelihood of dome closure due to sever dust at the MKO than at the ORM. I draw attention to this because ever since I began astronomical site testing on the peaks of Tenerife in the sixties, as mentioned in the earlier chapters, the question of Saharan dust has been magnified out of all proportion in order systematically to denigrate the Canarian Observatories.

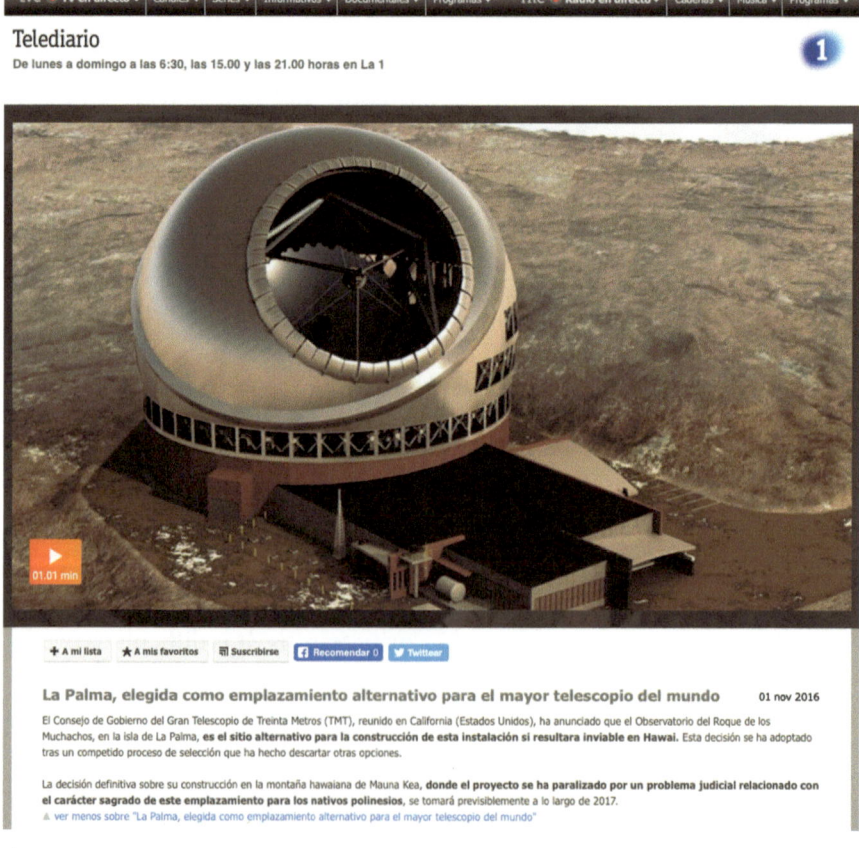

Fig. 22.4 Page of the RTVE website reporting on the decision adopted regarding the La Palma as an alternative site for the emplacement of the TMT (Thirty Meter Telescope). (Credit: RTVE)

In October 2018, during the commemorative events to mark the thirtieth anniversary of the Sky Law (see Chap. 14), the TMT's head of operations, Christophe Dumas, insisted:

> Roque de los Muchachos Observatory is fully capable of meeting the scientific objectives of the TMT, given the characteristics of the atmosphere at visible and near-infrared wavelengths, which are similar to those in Hawaii. Furthermore, the atmosphere above La Palma is stabilized by the ocean; it is a clean atmosphere without turbulence, above the sea of clouds. Most importantly, this Observatory already has the necessary infrastructure, and that greatly reduces the costs.

After all that has been said here, readers might be curious to know what the causes were that led to Spain losing Europe's giant telescope. If it was not for scientific, economic, operational, or security reasons, they would have to look for more confused motives, perhaps Spain's lack of interest in basic science, perhaps in vested interests or personal quarrels, or perhaps underlying fears. In other words, matters concerning human nature. This is not the place for me to enter into such a complex analysis. I am simply relating some of the facts for readers to draw their own conclusions.

On record and noteworthy are the efforts of the Chilean government to take very serious steps to get the E-ELT located in their country. Not only did they set up a high-level working group for the purpose, headed by the President of the Republic himself, along with all the ministers concerned in the matter, but they also took direct action in the member state countries of ESO through their ambassadors. After the tragic earthquake of 27 February 2010, alarm bells were sounded and the government initiated a diplomatic offensive. What I am relating here was reported chapter and verse in epic style in the Chilean newspaper *El Mercurio* on 2 May 2010. To complicate matters yet further, events coincided with a change of government. On taking over from Bachelet, Piñera immediately took the reins and went to see President Lula and managed to get Brazil to make known that it wished to join ESO and was willing to meet the shortfall for the construction of the E-ELT, the same amount of 300,000,000 euros that Spain had put on the table to ensure that the telescope would go to La Palma. Incidentally, Brazil still has not kept its promise and there are no signs that it is going to join ESO. The newspaper stressed that they were very worried in Chile because, 'Spain had the advantage of being a member of ESO, which gave it a voice inside the organization'.

The contrast with Spain was sobering. The apathy with which the government of Mr Zapatero treated the question of acquiring the giant European telescope for Spain, despite the economic advantages and prestige that this would bring, was incredible. It is astonishing, to say the least, how a deaf ear was turned to the clamour from the Canaries and the repeated demands of the opposition in Madrid. Zapatero did not even heed the unanimous petition of Parliament, including his own socialists, to make use of Spain's presidency of the European Union to push for actions to be taken to get the E-ELT located in the Canaries. There were many opportunities to do so, given the numerous meetings of both the President and his ministers in Europe.

One should not forget, as I have mentioned before, that ESO is a singular and powerful scientific and technical organization that is well entrenched within the European establishment, just like CERN, the European organization for nuclear research, or ESA. They are exclusive, expensive, and highly bureaucratized organisms that need always to be in the public eye with spectacular achievements. The highly paid upper echelons of their governing and managerial bodies are resistant to change and competition. This reality played a far from trivial role in the manoeuvring that led to the E-ELT being sent to Chile.

Let us now turn to Spain. Mrs Cristina Garmendia headed the Ministry of Science, Technology, and Innovation from the spring of 2008. As soon as she took possession of the portfolio, I informed her of everything relating to the IAC, whose Governing Council she would be chairing as one of her duties. I spoke to her of the E-ELT in particular and of the great possibilities and advantages of having it built at the ORM. I ended by asking for her help politically and economically. Not thinking twice, she replied that she would not support the project until she was certain that 'all Europe's astrophysicists were of unanimous accord that the telescope should go to the Canaries'. She could not, she said, permit yet another slip-up, by which she meant Spain's loss of the Source of Neutrons by Spallation for Bilbao, and the subsequent campaign by Basque news media and political parties throwing in her face her previous public promises that such a thing would not occur. No politician wants to back a losing horse. Once bitten, twice shy.

The growing suspicion on the part of the Minister that Spain's candidacy would fail became a certainty, and the reason for this must be sought in the advice of certain astrophysicists in her confidence. Among these there come to my mind the President of CSIC (then Rafael Rodrigo) and Xavier Barcons, then coordinator of the National Programme for Astronomy and Astrophysics. Both held CSIC research professorships. I understand, from what they said in private and in public, that they thought ESO would win. They were admirers of the great power of ESO and had first-hand knowledge that that organization wanted the telescope in Chile at all costs. Their declarations defending the advantages of Chile in the news media when the subject was aired in public were most revealing, so much so, that their words might have been dictated by ESO. One would then see Mrs Garmendia use similar words in Parliamentary exchanges and the press.

I sent the Minister three letters informing her on these matters. I explained the hidden motives behind ESO's closing of ranks in the question of putting the E-ELT in Chile and indicated the only way to upset their manoeuvres. I told her that it was a matter of urgency to show the ministerial departments of the member countries, who were paying hefty annual quotas and subsidizing the cash outflows of ESO, the figures revealing the substantial savings that would be made by putting the E-ELT in the Canaries. Not only would this result in lower construction and running costs but funding could also be obtained from the European Commission on the grounds that the Canaries are classed as ultra-peripheral territory, for which specific subsidies existed. The savings were so overwhelming that this would be a crushing argument to make. Remember that all this was going on during Spain's presidency of the European Union, and our ministers, secretaries of state, and others met frequently with their homologues in the other member states paying for ESO. I ended by asking that state diplomatic actions be taken, as was happening in Chile. I received the following reply from the director of her office, scolding me for my frankness and contacting the Minister directly: 'I wish you to understand that the Ministry would not permit its civil servants or high officials to repeat your affirmations, which might harm our institutional relation with the intergovernmental organization ESO'. It was shocking that the Director of the IAC should receive such a rap on the knuckles for drawing attention, as was his duty, to rude reality from the President of his own Governing Council, who was also the Minister of science. This correspondence is kept in the IAC archives.

In the end, the IAC, and the people and politicians of the Canaries, were the only ones who defended 'against wind and tide' Spain's candidacy for the E-ELT. In the local parliaments, the press, and everywhere we pressed home the exceptional benefits and advantages of the Canaries. We highlighted our strengths: the astronomical quality of our summits, protected by law; the lack of seismic activity and strong winds, allowing the possibilities of giant telescopes to be taken to their limit, thereby simplifying their design, construction, and running; the reduction in construction and running costs, etc. Added to all this was the advantage of the high standard of living in this territory of European culture. Naturally, we also preached the multiple benefits that hosting the E-ELT would bring to Spanish science and technology, and to the economy. They were crowded years.

It was encouraging to feel the popular warmth and backing of the Canarians in our struggle. As an example of this, I mention only one curious incident that I experienced personally. In the Plaza del Adelantado in La Laguna, there is an old cloistered convent in which the mummified remains of a nun who died in an aura of sanctity are venerated. On one day of every month these remains are on display to the public; it is a very popular event, and there are always poor people at the doors of the convent's church. I was walking through the square one day and a beggar who was unknown to me suddenly blurted out, 'Don Francisco, they've robbed us of the super telescope through the fault of the ******* politicians in Madrid. I want you to know that we are all very sorry'. This unexpected consolation both took me aback and moved me.

Perhaps the only positive thing Spain has obtained from this useless war with ESO was that a Spaniard, Xavier Barcons, had managed to become first the President of the Council of ESO in 2012 and later, at the end of 2016, Director General of that organization, a very important post.

To bring this chapter to a close, I again insist with deep sadness that, had the European Extremely Large Telescope been brought to the ORM, it would now be almost finished, thus gaining an advantage of fifteen to twenty years over our Americans competitors, and with an overall saving in its construction, commissioning, and running of not less than 50 per cent. Will ESO ever give an explanation for these realities? Will it one day reveal the real reasons why they chose Armazones, when it has been clearly demonstrated that the conditions for astronomical observing at Roque de los Muchachos are better?

It serves no purpose to lament what might have been, but I have felt obliged to explain all that happened in a truthful way. I have agonized over whether or not to reveal all, and had I still been Director of the IAC, I might have prudently desisted. Nevertheless, at my time of life I feel free to lay the truth on the table. I also believe it to be a just recognition of the efforts of the many people, Canarians included, who put their hearts and much hard work into this beautiful enterprise only to see it frustrated.

23

The Starlight Foundation: A Step Beyond Astrotourism (2008–Present)

The communication, publishing, and outreach activities of the IAC have gradually increased and diversified to the point of overload (see Chap. 20). Ideas and projects have had to be postponed indefinitely. By the end of the nineties there were ten people working in these areas in the Director's Support Team. To our lack of resources was the added problem of bureaucratic red tape. We therefore considered outsourcing the greater part of these activities, and the model that seemed best suited to us was that of a foundation.

There is a series of reports and decisions taken in this regard in the minutes of the meeting of the Governing Council of the IAC, dating from the first years of this century, in which the idea of a private foundation is accepted 'to ensure that the IAC, through its objectives, statutes, and Council, is guaranteed control over it'. There were a number of conditions attached: that the foundation's objectives included 'the spreading and popularization of Science, especially Astrophysics'; that it be 'an instrument through which the citizenry of the islands should achieve increased cultural and economic returns'; and that 'it serve generally to channel and facilitate the IAC's relations with local society'.

After making enquiries, putting out feelers here and there, and many formalities, we contracted a consultancy firm to produce a project for a foundation, together with an economic feasibility study. What we got from them in 2008 was a gargantuan organization, complete with bells and whistles, that was totally unviable, when our approach had been realistic and minimalist. And there the matter rested. But, as is our wont, we did not stop there. We continued to seek workable solutions to surmount the problems of lack of resources and flexibility. We were firmly convinced of the need to do science outreach and at the same time make it known that the IAC was working for the benefit of all.

It was at that time that a serious problem came to light in La Palma: a campaign had been mounted against Roque de los Muchachos Observatory. I explain in the course of this chapter how we finally found a solution.

In 1988 the Law on the Protection of the Astronomical Quality of the IAC Observatories, popularly known as the Sky Law, provided legal protection of the Canarian Observatories from light, industrial, and radioelectric contamination, and also prevented flights over the Observatories. The purpose of the law was to preserve the skies above the summits of the islands of Tenerife and La Palma as an international astronomical reserve (see Chap. 18). But we were coming up against too many difficulties in the application of the law in La Palma, in spite of the fact that the law was accompanied by economic aid for the reconversion of lighting fixtures.

It is interesting to see how the problem had arisen through certain vested interests playing on popular sentiment. It is not my intention here to mount a sociological study of the matter, but it is basically the same phenomenon that fans the flames of all kinds of separatism. I shall just give a few items of information.

A local ecologist group, wishing to make a name for itself and accustomed to being subsidized to keep it from causing trouble, had set its sights on Roque de los Muchachos Observatory, the giant infrastructure being set up with generous international resources. It was not difficult for them to persuade the population that up there on the mountain the boffins were producing dangerous waste that was contaminating the island's aquifers. If 'they' were dumping hazardous substances, the ecologists argued, all that poison would find its way into the water galleries from which the island's potable water was extracted. There was waste from toilets, they said, and enormous scientific instruments were spewing out harmful materials. That was why, they claimed, cancer levels were rising on the island. They even went so far as to assert that there were military lasers and other secret weapons in the Observatory. It is an old ploy: if you are up against a well-known entity, run to the news media and set yourself up as a defender of the people, a simple and inexpensive way to earn fame and cachet.

Then advantage was taken by certain opportunistic politicians to gain votes by following this populist route, fanning rumours, and declaring, 'They won't let us see during the night' with those 'dim yellow lamps'; they are 'robbing us' and 'using our summits, giving us nothing in return'. In this way they managed to foment a general rejection of the ORM among the population of La Palma that became deep-rooted and entrenched.

The short-sightedness, narrow-mindedness, and egotism of some of our politicians, and the manner in which they systematically deceive and manipulate the public to stay in power are very worrying. In the case that we are concerned with here, it beggars the imagination that such important public institutions as the Island Cabildo, being in possession of several independent and concordant rigorous scientific and technical studies demonstrating the innocuousness of ORM waste products, should continue shamelessly to deceive their citizens. They were also aware of the notably higher visibility of the island and the increase in tourism as a consequence of the ORM's prestige. But no matter. One Cabildo president even went so far as to demand that my title of Adoptive Son of the Island be withdrawn because of the harm I was doing to La Palma. The ecologists denounced me for committing an ecological crime. It all got to the point at which the presence of the IAC's observatory at Roque de los Muchachos was considered on the island as a burden and obstacle to its development.

Such, amazingly, was the panorama in La Palma at the beginning of this century, so Luis Martínez and I decided to take a conceptual leap forward and aim our sights high. We organized on the island a big international conference to roll back this tide of negativity. We sought the help of an expert in such matters: Cipriano Marín. I had met this strange person during the World Conference on Sustainable Tourism, which he had organized in Lanzarote in 1995, where I had given a talk on 'Astronomy: gathering experience for cultural tourism in the Canaries'. Together, the three of us planned the Starlight International Conference in La Palma in 2007. We were careful in its preparation and tried to secure the participation of people of great importance. Cipriano soon managed to arrange the participation of the International Astronomical Union, UNESCO, the Spanish Government, the Canarian Government, the Cabildo of La Palma, the World Tourism Organization, the European Parliament, the greater part of the UN conventions and programmes relating to nature conservation and the environment (CBD, CMS-PNUMA, the Ramsar Convention), the MaB Programme, the European Landscape Convention, and institutions and organizations from over thirty countries.

The meeting was convened under the title *A Common Heritage: An International Conference in Defence of the Quality of the Night Sky and the Right to Observe Stars*. Its formal objective was to promote a 'universal declaration of the right to starlight', as an expression of the commitment to protect from threat the common patrimony of humanity. But our underlying intention was to make a statement in La Palma so that its authorities and citizens should realize that we were not contaminating anything, and that they should appreciated the visibility and importance that their island was gaining thanks to astronomy.

A number of important experts in astrophysics, biology, the environment, lighting, tourism, education, and even lawyers met in La Palma in 2007. There were more than 200 participants. The President of the Canarian Government, the Minister for the Environment, the Vice-president of the European Parliament, the President of the Biosphere Reserve of La Palma, the Director of the Division of Ecological and Earth Sciences of UNESCO, and I presided over the inaugural session.

Animated, productive, and heated discussions and working groups were formed. A document approved by the conference was drafted in Spanish and English, and was distributed worldwide as the 'Declaration in Defence of the Night Sky and the Right to Starlight'. It was presented in a solemn public ceremony in the patio of the San Francisco Convent (converted since the IAC inaugurations of 1985 into a museum and cultural centre). It was a huge success.

The IAC organized a series of parallel popular astronomical events in various municipalities on the island, with conferences and even a touring *Cosmoneta* van of the Museum of Science and the Cosmos. There were also two exhibitions in Santa Cruz de la Palma. One, 'Astronomy for All Publics', demonstrating the Gran Telescopio Canarias, whose first light was scheduled for the following summer, was presided over by the King and Queen of Spain in the San Francisco Convent in Santa Cruz de La Palma. There were models and large-format photographs of both the telescope and Roque de los Muchachos Observatory. The other exhibition, *Cosmocolor* was held at Salazar Palace and exhibited the best photographs of the sky obtained by amateurs who had participated in the *Fotocósmica* contest, also organized by the IAC.

To hammer the nail home, in 2009 we organized in La Palma an international meeting of experts in the patrimony of humanity, intelligent lighting, nocturnal sky protection, Starlight Reserves, and star tourism. The title of the meeting was 'Where the Earth Meets the Universe'. It was another success with wide technical and media coverage.

The 'Declaration in Defence of the Night Sky and the Right to Starlight' began with a series of considerations and principles based on socio–cultural realities and international resolutions, such as the United Nations Universal Declaration of the Human Rights of Future Generations, which says: 'Persons belonging to future generations have the right to an unspoilt and uncontaminated earth, including the right to a clear sky'.

I cannot resist mentioning that the phrase 'including the right to a clean sky' is the result of a discussion I had many years ago with Commander

Cousteau himself, who opposed its inclusion when we were both members of a UNESCO panel of experts that prepared the Declaration.

Returning to the La Palma Declaration, attention must be drawn to its claim for the 'right to an uncontaminated night sky that allows the firmament to be seen and observed so that all may enjoy its contemplation', this right being 'equivalent to all the other environmental, social, and cultural rights'.

Consequently, there was an insistence on 'the control of light contamination', not only to guarantee astronomical observing but also to preserve nocturnal ecosystems, and all of this for biological, sanitary, scientific, cultural, and even energy-saving reasons. The aim was to make a definitive contribution towards creating a planetary consciousness and urge humans (who are all astronauts on this planet called Earth) to reach a supportive planetary commitment to preserve our Earth from the grave damage that we ourselves are inflicting upon it.

Emphasis was of course laid on 'protecting the skies in places of extraordinary astronomical quality', with suitable legal regulations, similarly to the Sky Law that safeguards the IAC Observatories. Astronomical tourism was also emphasized as cultural tourism and a source of rural development.

The Declaration ended with a series of concrete recommendations and petitions to international organizations, governments, local authorities, scientific institutions, and organizations involved in the protection of the environment, cultural diversity, and the promotion of sustainable development.

Ten years later, in April 2017, we organized, again in La Palma, but not without considerable effort, another international conference to drum up support for and update the 2007 Starlight Declaration. The title of the conference was *Preserving the Skies: 10th Anniversary of the Starlight Declaration*. Under the honorary presidency of King Felipe VI, there was a very high and significant representation of persons and international organisms from the world of science, conservationism, and tourism.

During meetings of groups of experts, it was established that the achievements of the application of the Starlight Declaration, although many, had not been able to prevent the significant build-up of planetwide light contamination closely connected with development and the growing concentration of city populations. Also analysed was the effect that the LED (light-emitting diode) revolution was having. Their evident energy efficiency needed to be counterbalanced against the risk of collateral damage to health posed by these light sources. Abundant data from recent scientific publications were used in these discussions. Attention was also drawn to the harm produced by some LEDs on the environment.

Fig. 23.1 The renowned oceanographer Commander Jacques Cousteau conversing with the author at the meeting of experts convened by UNESCO to discuss the 'Universal Declaration concerning the Rights of Future Generations' in 1994. The Declaration was later approved by the United Nations. (Credit: IAC)

There was general agreement that the key to ameliorating all these problems lay in the basic general education of children and adolescents, and therefore in the sensitivity and knowledge of teachers and educators.

The final document, published in English and Spanish was a good one, but it had been difficulty to draft it to everybody's satisfaction. It was clear in its ideas and splendid in its wording. It is well worth reading and can be accessed on the internet by searching for 'Visions and Resolutions: 10th Anniversary of the Starlight Declaration'.

After an introduction on the lessons learnt in the previous decade, there followed a call to action, with very specific analyses and recommendations ordered by chapters:

1. Recovery of the cultural values of the starry night
2. Preservation of biodiversity and human welfare
3. Towards a new illumination culture
4. The universal value of clear and dark skies for astronomy
5. Starry skies as drivers for sustainable development
6. Appendix with technical data concerning the limits for the different parameters

Returning to 2007, direct consequences of that conference and of the La Palma Declarations issued therefrom that must be emphasized are the launching of the Starlight Initiative, the formal establishment of the Starlight Foundation, and above all the drastic change in the perception of the Roque de los Muchachos Observatory by the people and authorities of La Palma.

The Starlight Initiative is conceived as an international campaign in defence of the quality of the night sky and the right of all to be able to contemplate the stars. It calls on all scientific, cultural, and citizens' institutions and associations to join forces with it in its goal to demonstrate the importance that clear skies have for humanity by highlighting and making known the value that this heritage, now under threat, has for science, education, culture, tourism, history, and quality of life.

More than a hundred town halls and public administrations from all over the world have signed the La Palma Declaration in defence of the night sky and the right to starlight, with all that that entails in terms of commitment to protect and willingness to act accordingly—all this with free promotion of the island of La Palma, it must be added.

The La Palma Declaration has certainly received worldwide distribution and has served to support the many movements that have arisen to protect the night sky from light pollution. Some of these, such as the American Dark Sky Association (DSA), possess considerable strength and economic resources. Indeed, the DSA are competing with our Starlight Foundation with products analogous to our certifications. Given its power, it would not be surprising if it ended up achieving backing for its products from the international organizations that guarantee Starlight Certifications.

But the most important result, from a local perspective, of the international Starlight Conference of 2007 was finally to convince the citizens of La Palma that Roque de los Muchachos Observatory was a source of pride for them. They have at last come to understand that astrophysics has not only put their island on the map but is also a direct source of revenue. It is well established that astrophysical activity as a whole is now the most important and fast-moving enterprise on La Palma, and that the fame of the island's unpolluted skies, due precisely to the world-famous Roque de los Muchachos Observatory, has become a permanent lure for tourists. Astrophysics gives the island a unique, powerful, and very special identity.

We have been saying this since the sixties, asking the citizens and their authorities to prepare to exploit the new wealth that was coming their way. When I was interviewed regarding the signing of the multinational Cooperation Agreements for Astrophysics, I said that they should begin to promote La Palma as an 'Island of sun and stars'.

The Town Hall of Santa Cruz de la Palma and the Cabildo are now creating a Science Walk in the new seafront promenade with stars commemorating, Hollywood style, the names of famous scientists who promote the island through their observations at Roque de los Muchachos Observatory. I was very pleased to read in the press that the Nobel laureate Samuel Ting has been awarded one of these stars. In the words of the current President of the Cabildo: 'It is a source of great pride to see the wide consensus and support that many kinds of institutions and the public at large in declaring La Palma to be the island of the stars. It is a well-earned and established trade mark that distinguishes us as a singular destination in a regional, national, and international context'.

It has been a long and arduous journey. More than thirty years have passed, but the reward is a welcome one.

I have already mentioned, with reference to the Starlight Conference of 2007 and the La Palma Declaration, that these have helped us in finally establishing the Starlight Foundation for the popularization of astronomy. In 2009 the Foundation was presented for registration and delivered its statutes to the Canarian Registry for Foundations. But its registration was not completed until 2011, a mere two years of bureaucracy—not bad going!

The Starlight Foundation was established to popularize astronomy and manage, boost, and disseminate the philosophy and principles contained in the La Palma Declaration. It was created by the Instituto de Astrofísica de Canarias and a firm called Corporación 5 (C5), with a majority holding by the latter, precisely so that its legal character would that of be a private company, thereby avoiding bureaucratic red tape. Specifically, the IAC put up 5000 euros and C5, 6000 euros. The creation of a foundation with a capital less than 30,000 euros is not normally permitted, and it was difficult to convince the authorities that we would make up the shortfall in the Foundation's social capital in the future, this being one of the reasons behind the formal establishment of the Foundation.

The activities pursued or promoted by the Foundation are fundamentally based on the spreading of scientific knowledge in general culture through astronomy and the protection of the starry sky as a top level scientific, cultural, environmental, and touristic treasure. It strives to ensure that uncontaminated skies are also seen as a natural resource for the economic development of depressed regions. As a matter of fact, astrotourism is on the rise and presupposes intelligent illumination, energy saving, and the protection of many nocturnal species.

A basic tool for fulfilling these objectives is the international system of certifications created by the Foundation with the backing of UNESCO, the

International Astronomical Union, and the World Tourism Organization. Starlight certifications accredit sites as Starlight Touristic Destinations, Starlight Reserves, and so forth. These accreditations are awarded following a stringent protocol, an essential prerequisite being the acquisition and confirmation of data on the astronomical quality of the dark sky of the site concerned through audits carried out by independent experts. These are a solid guarantee that at sites so accredited the night sky may be observed and enjoyed.

Starlight Touristic Destinations are accessible territories that enjoy excellent conditions for observing the night sky and, being protected from light pollution, are particularly apt for the development of touristic activities based on that natural resource.

To receive their certifications, these spaces must not only prove, with measurements made in situ, the quality of their skies but must also guarantee their protection, have adequate infrastructures, and personnel trained to conduct groups during the night, and observe and interpret the firmament, thus spreading astronomical knowledge. Later, independent auditors pay a visit, confirm the data provided, and send a report to the Foundation with their assessment and any recommendations that they see fit to make. Finally, if all is in order, the Foundation issues the corresponding Starlight certification.

Starlight Reserves are territories which, independently of the protection they already enjoy as biosphere reserves, natural parks, etc., have the additional qualification and guarantee, through this certification, of their skies. This allows them to incorporate enjoyment of the night sky into their other cultural, scientific, landscape, and natural resources.

In order for the auditors to take action they must adopt a series of reference parameters for astronomical quality with their corresponding limiting values. These were defined at a meeting of experts at the headquarters of the World Tourism Organization of the United Nations. Of decisive importance was the expertise of the IAC's Sky Quality Group, led by Dr Casiana Muñoz and of which Dr Toñi Varela also forms part. It was the increasing involvement of Dr Varela in Starlight certifications, combined with her good nature and cordiality, that has won her great popularity among those who manage tourist destinations and Starlight reserves.

The Starlight Foundation has extended its activities to rural houses, hotels, paradors, camping sites, star parks, stellaria, and other establishments which, besides formally adhering to the principles expressed in the La Palma Declaration, offer their clients astronomy outreach activities and facilities for observing the sky.

Fig. 23.2 Monsaraz Convent in Alqueva (Portugal), a certified Starlight Touristic Destination. (Credit: Miguel Claro)

Fig. 23.3 A practical session during a Starlight Astronomical Monitor course at DTS Gredos Norte. (Credit: José Jiménez)

In order to make all this a feasible proposition it was necessary to begin training Starlight auditors, as well as guides and Starlight monitors. Specific training courses for Starlight Auditors have been given to enable auditors to examine technically both the sky quality and the conditions and settings of the place where the corresponding activities are to be developed.

For people who have covered only a part of the content of these courses there are shorter courses for them to complete their training, especially in astronomy. These are awarded the title of Starlight Monitor.

To get an idea of what has been done about training, the following figures for people with Starlight training are for the year 2018: 49 guides, 96 astronomical monitors, 18 auditors, 16 school teachers, and 30 journalists.

Particularly noteworthy is the huge impact of what the Starlight Foundation has been doing on the Spanish news media. In 2017 alone the Foundation appeared 350 times on television and in print.

Readers who wish to know more about the Foundation and its many activities in tourist destinations, reserves, and other associations will find further information in its attractive and very complete website: www.fundacionstarlight.org. There they will find, in Spanish and English, everything from a gallery of beautiful images and astronomical videos, and very full information on the past and present activities of the Foundation to news and business opportunities.

All this complex and specialized activity that I have just described, together with the notable results that are being obtained, is being achieved on a shoestring budget. It needs to be pointed out that the Starlight Foundation has from its creation been operating in a condition of crippling privation that is

Fig. 23.4 Astrotourism activities at the La Palma Starlight Touristic Destination. (Credit: Antonio González)

holding it back from realizing its great potential. It started in 2009 with a legacy of 11,000 euros, a sum that was increased by 6000 euros put up by Corporación 5. Another 5000 euros that should have been contributed by the IAC was not forthcoming. In the absence of a wealthy patron, the Foundation gets by from the little income it receives for certifications, which barely covers the cost of processing the documentation involved. It has to be said that the Foundation has sunk to such an extreme level of penury that it has sometimes had to resort to taking out a loan to pay the paltry salaries of its two permanent employees. Its Director, the soul of the Foundation, Luis Martínez, an expert in communication (Chap. 20) has no salary and regards his work as a labour of love, so all the gains described earlier are also the result of personal commitment.

The cultural—not the economic—success of the Starlight Foundation reminds me of what occasionally happens with low budget films, which can sometimes achieve a quality and fame denied many lavishly funded superproductions. This causes me to reflect on patronage in Spain and other matters. There has been no way to get any sponsorship. Neither have we been able, or known how, to set up a company capable of competing in the international market for trademarks required by sustainable cultural tourism. It seems we are not cut out to make money.

24

Astrophysics in Spain: The Wider Picture (1974–Present)

One day somebody ought to write the complete history of the birth and establishment of astrophysics in Spain, an event that occurred in the second half of the last century. I do not have the energy to tackle the task myself. All I have done is to mention facts here and there throughout this book, particularly in this chapter. I have tried to relate matters that, in my judgement, were of importance in detonating the explosion of astrophysics in Spain. This burst of activity was not merely a consequence of the arrival of telescopes at our summits from overseas. Always behind such opportunities there have been people, pitching in with their passion and intelligence, who knew, or sometimes did not know, how to make the best use of them.

What happened was extraordinary and the events overwhelming. In half a century we have risen from practically nothing to joining the most scientifically advanced countries in the world. In the first chapter I explained how, newly married, full of ignorance, but with great enthusiasm, I landed, without a parachute on the summits of the island of Tenerife. Astrophysics had not yet arrived in Spain. But today's panorama is very different. Spanish astrophysicists have risen in number from a mere handful to almost a thousand. There has been an avalanche of new astrophysics researchers. The most complete and recent data are to be found in reports of the National Commission of Astronomy (CNA) and the Spanish Astronomical Society (SEA). These data tell us that Spain now occupies seventh place in the world with respect to scientific production in sciences of the Universe (Scimago, 1996–2016). In other words, we are therefore situated behind the United States, Germany, the United Kingdom, France, Italy, and Russia, and we are ahead of countries such as Japan, China, Australia, and Canada.

F. Sánchez, *The Rise of Astrophysics in Modern Spain*, Astronomers' Universe, https://doi.org/10.1007/978-3-030-66426-8_24

It gives me great satisfaction to see articles authored by Spanish research workers in every issue of the foremost astronomical journals; contrast that with the situation in the mid-sixties, when only one humble astrophysics article of mine was to be found in a refereed journal.

But that is not all. At the beginning of 2018 the Economics, Accountancy, and Finance Department of the University of La Laguna published a detailed study which stated that, for every euro spent on astrophysical activity in the Canary Islands in payment for goods and services, 3.56 euros were generated for the GDP of the archipelago; furthermore, the report went on, for every million euros 45 full-time jobs were generated annually (see Chap. 15).

It would most instructive if our economists and business management experts were to analyse this 'miracle' of Spanish astrophysics from their level-headed and rigorous perspective. Its rise from the humblest of beginnings to today's countless successes might usefully be compared to that economic phenomenon of the company ZARA-INDITEX (although, of course, there are evident differences). I mention this because I believe that, for Spain to achieve its rightful place in science, our businesses and financial institutions need to be aware of such facts as the role of astrophysics in Spain. In a more developed country, where scientific research would be regarded as a real investment, such a study would have been carried out long ago. Nevertheless, dreaming on, such a study might also serve to encourage our politicians to make a genuine effort to boost Spain's much-trumpeted RDi.

By the mid-twentieth century, progress in astronomy and notable advances in telescope construction, themselves a consequence of accelerated technological development brought about by the Second World War, revealed the need to find suitable sites for the new telescopes. Traditional observatories did not meet the stringent new requirements for astrophysical observing, and European astronomers turned their thoughts to sunny Spain. Thus originated Calar Alto Observatory in Almeria, and Teide Observatory and Roque de los Muchachos Observatory in the Canaries.

This international interest in the skies of Spain for the purpose of setting up modern telescopes caused Franco's dictatorship, which badly needed to establish international agreements, to be more than willing to conform. At the time it was rumoured that the substantial infrastructure needed in the Filabres Sierra in order to access the German observatory in Calar Alto was built in exchange for the importation by Germany of Valencian oranges. This did not, of course, figure in the agreement signed by the National Commission of Astronomy and the Max Planck Institute. Although named the Hispano–German Observatory, it was in reality a purely German enclave.

Fig. 24.1 One of the telescopes of the Calar Alto Observatory (Almería). (Credit: DigiGalos)

The attitude of our European colleagues was the usual one of that time. It had been applied in Chile in the building of the European Southern Observatory, the former colonial Royal Observatory of the Cape of Good Hope (now the national South African Astronomical Observatory). The owner institutions would build their own observatories, make them as self-sufficient as possible while hiring locals strictly for cleaning and auxiliary services. Overseas observatories were mere observing outposts, all the science and technology being done in the home country.

The key to breaking with this concept in Spain was understanding that the exceptional astronomical quality of the skies above our mountain summits was a 'public natural resource' that needed to be exploited for the benefit of our own science and technology—in other words, our country. All this is recounted in Chaps. 6 and 10. Right from the start we were clear—from a global, generous, and national point of view—that the astronomical skies of the Canaries were the property of all Spaniards.

Another important step was immediately to commence training Spanish astrophysicists with the help of our colleagues from all over the world (see Chap. 8). Our young researchers went abroad to learn about modern astronomy and returned purged of all inferiority complexes and ready to do research here.

One of our earliest successes in following this path in 1975 was to pioneer a 'national plan for the training of astrophysicists'. National plans were all the rage at the time. It was a huge triumph, for a good many of the first outstanding researchers and university lecturers in this branch of science began their training and acquired their skills as grant holders under this national plan.

We could see that, even though the large observing instruments had to be concentrated in a few select sites, the actual astrophysical research needed to be spread throughout the country. It was becoming daily less necessary to be standing under a telescope to do science; it was important to encourage and promote the creation of astrophysical research teams at universities and other institutions in the country. I repeated this endlessly to ministers, university chancellors, and anybody else who cared to listen to me.

To give an example, I mention an incident that led Minister Cruz Martínez Esteruelas, after his visit to Teide Observatory in 1974, to believe in us and in the future of astrophysics in Spain. On returning to Madrid, he asked the then President of CSIC, Professor Primo Yúfera, to come and visit us. Yúfera duly made his visit and was converted to our cause as soon as he saw the reality of what we were doing and understood what the future might hold. I must say that this head of CSIC and Alejandro Nieto who followed him have been the only heads of that institution who have truly understood and supported us.

We encouraged Primo Yúfera to follow our model and create an institute of astrophysics as close as possible to the Hispano–German Calar Alto Observatory in Almeria with the aim of gradually introducing a Spanish flavour to this German enclave. In the early days of Calar Alto Observatory, it was forbidden to raise the Spanish flag, and Spaniards were employed purely as service staff. I used to refer to it as a mini-Gibraltar to ram home the point. In May 1974 my name was put forward as Spanish director of this observatory, but I rejected the offer because I wanted to devote all my efforts to the Canaries. That is how, in 1975, the Instituto de Astrofísica de Andalucía (IAA) was created by CSIC with its headquarters first in Almeria before being transferred shortly afterwards to Granada. The name is clearly in the same style as that of the IAC but for Andalucía. It gives me great joy to see that the IAA is today a first-class centre of research and technological development. Its first Director, and its driving force, was José María Quintana.

While on the subject of the Hispano–German Calar Alto Observatory, I must say that it was quite a struggle to get the recently built 1.5-m optical–infrared telescope of the Royal Observatory of Madrid (ROM) installed there. First of all, the ROM astronomers wanted it set up in Yebes so that it would be within easy reach. Added to that, the Germans did not want Spanish telescopes in 'their' observatory.

International agreements were signed in the sixties to permit the most advanced telescopes of the day to be set up in Spain at Izaña, Roque de los Muchachos, Calar Alto, and Veleta. The economic boom began to be felt in the country, as was also evidenced in research investment with the creation of new centres and jobs.

It was also in the sixties when modern astronomy began to find a home in Spain. The first embryonic research teams were born during these years, and they were formed around certain figures. Here is a list of the most notable of those pioneers in the order in which they made their appearance:

- Canary Islands: Francisco Sánchez, Juan Casanovas, Carlos Sánchez
- Granada: Eduardo Battaner, José María Quintana, Ángel Rollán, Mariano Moles
- Barcelona: Juan José de Orús, Ramón Canal
- Madrid: Manuel Rego, María José Fernández

I have listed here only the names of the very first people to become involved, so the numerous individuals that followed, and who made their own significant contributions to helping Spanish astrophysics occupy the position that it

Fig. 24.2 Sierra Nevada Observatory of the Instituto de Astrofísica de Andalucía. (Credit: IAA)

holds today, do not appear (in the interests of brevity and to avoid converting this roll call into a cumbrous tombstone).

Special mention must be made of radio astronomy in Spain, also in the sixties, which got off to a fine and brilliant start, thanks to the physicist Jesús Gómez, who trained as a radio astronomer in France and the United States. As soon as this pioneer passed his *oposición* for a post as astronomer at the Royal Observatory of Madrid he converted Yebes Observatory and its instrumentation into a competitive radio astronomy centre. He later reformed the ROM by introducing astrophysics into that archaic institution, endowing it with resources, and negotiating beneficial international agreements. He also dignified its historical headquarters in the Retiro Park (Madrid) by turning it into a science centre and museum. There is now a fine full-size reproduction of the largest telescope of the early nineteenth century, built by Herschel (the original never saw first light because it was burnt to the ground by Napoleon's troops).

To continue, many complicated battles had to be fought and won in order to break down the old structures of Spanish astronomy. So much talent was lost after the Civil War, Spain was ostracized and submerged in misery, and universities and research centres foundered while clinging on to the old structures. To make the best use of the modern telescopes that were now (in the seventies) on their way we needed new organizations better able to cope (see Chap. 15).

Another ancient relic was the National Commission of Astronomy (CNA), created by the government in 1920. It was the only official body that then existed in Spain to direct and coordinate astronomical activities. It was clearly necessary to renovate the Commission, given that it was governed by a decree issued in March 1948. among whose many archaisms was the Committee's composition, with representation solely from the old observatories of the past and the three chairs of classical astronomy. It took years to reform this. I used my good relations with Madrid to get the stale decree modified, and in 1989 Royal Decree 587/1989 regulating a now updated CNA was approved. The Commission was now established as a:

> Collegiate body of the State Administration whose aim was to direct and coordinate national astronomical programmes, and to represent Spain in the International Astronomical Union, carrying out its mission in accordance with directives determined by the Inter-ministerial Commission of Science and Technology . . . [and] in general undertake such activities as are related to the promotion, orientation, and scientific teaching of astronomy in Spain.

The CNA was substantially transformed with the entry of new membership comprising the following:

- The Director of the Esteban Terradas National Institute of Aerospace Technology
- The Director of the Royal Naval Institute and Observatory
- The Director of the Instituto de Astrofísica de Canarias
- The Director of the Instituto de Astrofísica de Andalucía
- The Spanish delegate of the European Space Agency
- The Deputy Director General of Astronomy of the National Geographical Institute
- One scientist of world standing for each of the following fields: Galaxies, Stellar Physics, the Interstellar Medium, the Sun and Solar System, Positional Astronomy, Instruments, and Space

However, the Director General of the National Geographical Institute (IGN) was to remain as President of the Commission, even though it was felt that that position should be held by an astronomer, duly elected by the Commission membership. Well, at least seven renowned scientists had been added to the membership.

This rule was later changed, the presidency now alternating between the Director General of the IGN and the President of CSIC. However, because the Commission has never had sufficient dedicated funding to enable it to perform its duties, its activities are of little consequence.

The few young Spanish astrophysicists of the seventies felt the need to get to know one another in order to make a joint push for research in their branch of science. Meetings among teams of colleagues took place and it was decided to take a step further in 1975 by organizing the First National Assembly of Astronomy and Astrophysics, to be held in the Canaries. It was a great success and is fondly remembered by those of us who attended. Together with the young astrophysicists, the older astronomers from the observatories in Madrid and San Fernando participated in the conference. The Professors of Astronomy Orús and Cid Palacios were there (Torroja did not come), and Father Romañá, president of the Alfonso X Board of CSIC gave a talk.

The flight to La Palma to see Roque del los Muchachos from the air (at the time there was no access road for a proper visit) was unforgettable. The plane we chartered was old, noisy, and creaky, and seemed on the point of disintegrating at any moment. The ventilation consisted of a lot of small fans fixed to the ceiling with wires. We all held our breath as the plane landed.

The assembly served the purpose of creating synergies and resulted in many contacts among research teams. These meetings became a regular and moveable fixture. Year after year we would meet in different cities to report on what we had been doing. They were the embryo of what would later become the Spanish Astronomical Society (SEA); it was during the IV Assembly, held in Santiago de Compostela in 1983, that the creation of the SEA was first mooted.

As a further demonstration of the growth of astrophysics in our country, in 1992 the Spanish Astronomical Society was founded. The main reason for its creation was 'to contribute towards the promotion and development of astronomy and astrophysics in Spain, and provide an independent forum for the discussion of matters of common interest for the astronomical community of Spain'. Finally, in 2001, the president of the SEA was made a voting member of the CNA, thus integrating both organizations.

I take pride in being one of the 39 founding members of the SEA and of the European Astronomical Society (EAS), created in 1990, of which I would eventually become Vice-president.

Because funding for research in these climes is almost exclusively public, it was mandatory to open channels for these resources to be created and duly

Fig. 24.3 Participants at the First National Assembly of Astronomy and Astrophysics on the stairway of the old building of the University of La Laguna in September 1975. (Credit: IAC)

Fig. 24.4 Participants at the thirteenth Scientific Assembly of the Spanish Astronomical Society in front of the monumental doors of the University of Salamanca in July 2018. (Credit: SEA)

labelled astronomy, astrophysics, space, and so forth. In 2000 we managed to get the National Programme of Astronomy and Astrophysics (PNAYA) explicitly included in the National Plans for RDi. There thus existed a stable framework for the consolidation of astrophysics in Spain. Until that time, it had been a matter of competing with everybody else within the General Promotion of Knowledge programme. As explained in Chap. 21, the construction and exploitation of the Gran Telescopio Canarias played a decisive role in all of this. The PNAYA has been a very useful tool for research teams dispersed throughout Spanish territory, as is well documented.

The following list demonstrates the result of this nationwide programme. This is how astronomical research is now distributed in Spain:

- Instituto de Astrofísica de Canarias (IAC)
- Instituto de Astrofísica de Andalucía (IAA)
- Astrobiology Centre (CAB)
- Institute of Space Sciences (ICE)
- Physics Institute of Cantabria (IFCA)
- National Astronomical Observatory (OAN)
- Centre of Energetic, Environmental, and Technological Research (CIEMAT)
- Aragon Centre of Cosmic Physical Studies (CEFCA)
- Institute of High Energy Physics (IFAE)
- Madrid Institute of Material Sciences
- San Fernando Royal Naval Observatory (ROA)
- University of Barcelona

- Complutense University of Madrid
- University of Valencia
- University of La Laguna
- Autonomous University of Madrid
- Polytechnic University of Catalonia
- University of Alcalá de Henares
- University of Granada
- University of Alicante
- University of Zaragoza
- University of the Basque Country
- Polytechnic University of Valencia
- University of the Balearic Islands
- University of Santiago de Compostela
- University Miguel Hernández
- University of La Coruña
- University of Salamanca
- University Rovira i Virgili
- University of Valladolid
- University of Cantabria
- University of Cádiz
- University of Oviedo
- University of Girona
- University of Vigo
- University of Huelva
- University of La Rioja
- University of Jaen
- University of Extremadura
- Polytechnic University of Cartagena
- University of Murcia
- National University of Distance Education (UNED)
- Other universities in Madrid (UPM, URJ, UEM)

With this panorama it seems clear that modern astronomy is now well established in Spain. It is unimportant where its centre of gravity is located at any given time or what its flagship might be. In cycling marathons those in the lead are relieved by others, and in migrating flocks of birds, different members of the flock occupy the apex. What is important is that we all pedal, or fly, together with spirit and enthusiasm for the common good of science.

Epilogue

You have now reached the end of the book and have read its chapters with—I hope—growing enjoyment and interest. If that is the case, then I may congratulate us both. Throughout the text I have striven to narrate the most important events in the short and fruitful story of astrophysics in Spain, and show how we have managed to reach such dizzying heights. I also wonder if the reasons behind what was done have been clearly enough stated. If I have succeeded in doing this, then I shall be satisfied at having contributed towards making known this most human aspect of science.

You have witnessed the pride with which I tell how the Instituto de Astrofísica de Canarias has become a competitive centre for research and technological development with a well-earned international reputation, its two internationalized Observatories acting as a magnet for advanced telescopes and observing instruments from all over the world; and how the IAC is also a postrgraduate school and outreach centre. To top it all, the IAC has designed, built and brought into operation the Gran Telescopio Canarias, thus entering the realm of Big Science and enabling national industry to compete successfully in the profitable technology associated with it. There are now many new paths to follow.

But that is not all. The role of the IAC in the birth and spectacular flowering of astrophysics in Spain continues to be a leading one. Let us not forget that in very few years Spain, starting from absolutely nothing, is now among the top seven countries in the forefront of this science (as explained in Chap. 24), with all the cultural, social, economic benefits implied for the country.

© The Author(s), under exclusive license to Springer Nature Switzerland AG 2021
F. Sánchez, *The Rise of Astrophysics in Modern Spain*, Astronomers' Universe,
https://doi.org/10.1007/978-3-030-66426-8

I cannot deny my joy and satisfaction at the constant flow of successes and triumphs achieved by the scientists and engineers of the IAC, and at seeing how highly valued are those who leave to work elsewhere. At this point in my life perhaps my greatest reward is witnessing how those who have succeeded me continue to reach ever greater heights. That is how the world progresses.

The IAC today is now firmly embedded and valued within its environment. The Canary islanders are proud of their *Astrofísico*, and their 'astronomical skies' are now part of the Canarian identity, a scientific referent—even to the point of becoming a tourist attraction. It would have been difficult to imagine all of this at the beginning, way back in the sixties.

The majority, if not the totality, of the recent spectacular advances in our astronomical knowledge are owed to instrumentation, to technology. Those giant prostheses—the telescopes, collectors of photons—continue to reveal to us hitherto unsuspected depths, confronting us with new, astounding, and wonderful mysteries. This is the future.

Astronomy continues to serve humanity with an ever-growing harvest of new knowledge in its quest to understand a little better the Universe in which we live, who we are, and where we come from. Consciously to participate in such a venture is the greatest reward to all who dedicate their lives to it.

I close by recalling that that which enabled us to be born, to get to where we are now, and make the IAC distinctive are the extraordinary and protected Canarian Observatories, the seat for the new photon collectors of the future. This redoubt, which we must continue to protect and promote, and the recruitment of the finest personnel are our best guarantee for the future. Let us not forget it!

Biblographical Notes

BEA: *Biographical Encyclopedia of Astronomers*
BOE: *Boletín Oficial del Estado*

Chapter 1

Perhaps It All Began with an Eclipse (1959–1961)

A general history of astronomy in Spain from its earliest beginnings until 2009 is provided in the doctoral thesis of Iván Fernández Pérez, *Aproximación histórica al desarrollo de la astronomía en España* (University of Santiago de Compostela, 2009). For an overview of the history of solar studies in Tenerife, see Manuel Vázquez Abeledo, *Observando el sol desde Tenerife: una Aventura sobre el mar de nubes* (La Laguna: IAC, 2019). Detailed descriptions of Spanish administrative and governmental structures during the Franco era are given by S. G. Payne, *Franco's Spain* (London: Routledge & Kegan Paul, 1968) and by J. Amodia, *Franco's Political Legacy: From Fascism to Façade Democracy* (London: Penguin Books, 1977). Severo Ochoa's acceptance address for the 1959 Nobel Prize in Physiology or Medicine is reproduced in *Nobel Lectures, Physiology or Medicine* (Amsterdam: Elsevier, 1964). The *Consejo Superior de Investigación y Ciencia* (*CSIC*, 'Upper Council for Science and Research') was created in 1939 by Law 24-11-1939 (*BOE 27 y 28-11-1939*). It replaced the *Junta para Ampliación de Estudios e Investigación* (*JAE*, 'Council for the Furtherance of Study and Research'). Teide Observatory (*Observatorio de El*

F. Sánchez, *The Rise of Astrophysics in Modern Spain*, Astronomers' Universe,
https://doi.org/10.1007/978-3-030-66426-8

Teide, later known as *Observatorio del Teide*), was created by Ministerial Order on 10 February 1959. The story of the author's role in developing astrophysics in the Canary Islands is told by José B. Navarro, *Francisco Sánchez Martínez: La Facultad de las Estrellas* (Centro Cultura Popular Canaria, 2018).

Chapter 3

Tenerife and Mount Teide on the Moon (1724–1914)

Ewen A. Whittaker, *Mapping and Naming the Moon* (Cambridge: Cambridge University Press, 1999), gives a detailed history of lunar nomenclature. Thomas Sprat describes an expedition to Pico Teneriffe ('A relation of the Pico Teneriffe') in *The History of the Royal-Society of London* (London: J. Martyn, 1667), 200–13. Van Langren is more commonly known to astronomers as Langrenus (see J. M. Lattis, *BEA*, vol. 1, 676–7). There are also entries for Riccioli (J. F. MacDonnell, *BEA*, vol. 2, 968–9), Schröter (C. J. Cunningham, *BEA*, vol. 2, 1030–1), and Birt (R. McGown, *BEA*, vol. 1, 131). For a useful description of scientific expeditions to the Canary Islands in the eighteenth century see Alfredo Herrero Pique, *Las Islas Canarias, escala científica en el Atlántico: viajeros y naturalistas en el siglo XVIII* (Madrid: Editorial Rueda, 1987). The Darwin quotation is from F. Darwin (ed.), *The Autobiography of Charles Darwin and Selected Letters* (New York: Dover Publications, 1958), 136.

Chapter 5

The Zodiacal Light (1962–1968)

A gentle, but now somewhat dated, introduction to interplanetary matter is given by F. G. Watson, *Between the Planets* (Cambridge, Mass.: Harvard University Press, 1956, 172–9). A rigorous treatment of dust grains is provided by L. Spitzer's standard text, *Physical Processes in the Interstellar Medium* (Cambridge: Harvard University Press, 1978) and H. C. van de Hulst, *Light Scattering by Small Particles*, 2nd edn (New York: Dover Publications, 1981). One of the first publications reporting work done at Teide Observatory concerned the photometry of the green flash: F. Sánchez and R. Dumont, 'Fotometría absoluta de la raya verde del continuo atmosférico en el

Observatorio Astronómico del Teide (Tenerife) de enero de 1964 a julio de 1965', *Revista de la Real Academia de Ciencias Exactas, Físicas y Naturales de Madrid*, LXI, 1 (1967). The author's doctoral thesis, *Contribución al conocimiento del medio interplanetario por fotometría y polarimetría de la luz zodiacal* (University of Madrid, 1969), gives a full account of work done on the zodiacal light with the Bordeaux Telescope. Other works covering ZL research at Teide Observatory include R. Dumont and F. Sánchez, 'Recent polarization measurements over the sky at Tenerife Island', in J. L. Weinberg (ed.), *The Zodiacal Light and the Interplanetary Medium* (NASA SP-150, 1967); R. Dumont and F. Sánchez, 'Photometrie de la lumière zodiacale hor de l'Eliptique en quadrature et en opposition avec le soleil', *Astronomy and Astrophysics*, 22, 321 (1973); R. Dumon and F. Sánchez, 'Zodiacal light photopolarimetry. I. Observations, reductions, disturbing phenomena, accuracy', *Astronomy and Astrophysics*, 38, 397 (1975); R. Dumont and F. Sánchez, 'Zodiacal light photopolarimetry. II. Gradients along the Ecliptic and the phase functions of interplanetary matter', *Astronomy and Astrophysics*, 38, 405 (1975); and R. Dumont and F. Sánchez, 'Zodiacal light photopolarimetry. III. All sky survey from Teide 1964–1975 with emphasis on off-Ecliptic features', *Astronomy and Astrophysics*, 51, 393 (1976). B. H. May's doctoral thesis, *A survey of radial velocities in the zodiacal dust cloud* (Imperial College, 2007) describes his work at Izaña using a Fabry–Perot interferometer and gives full references for past and present zodiacal light studies.

Chapter 6

Astronomical Site Testing (1961–1977)

The results of all the site-testing campaigns at the Canarian Observatories up to 1985 are summarized by F. Sánchez, 'Astronomy in the Canary Islands', *Vistas in Astronomy*, 28, 417. A more detailed account of the many site-testing campaigns at the Observatories is given in C. Muñoz-Tuñón (ed.), *Site Properties of the Canarian Observatories* (*New Astronomy Reviews*, 42, 395–572, 1998).

Chapter 7

Astrophysics Versus Astronomy (1961–1972)

Stanley Payne (*Franco's Spain*, London: Routledge & Kegan Paul, 1967), ch. 5, offers insights into the lamentable state of the sciences and arts in Spain after the Civil War. The origin of astrophysics is described by A. Pannekoek, *A History of Astronomy* (New York: Dover Publications, 1961, 395–418). Dieter B. Herrmann, *The History of Astronomy from Herschel to Hertzsprung* (Cambridge: Cambridge University Press, 1984), ch. 2, describes the extreme reticence of the European astronomical community in accepting the validity of applying physical methods to the study of celestial bodies in the nineteenth century. The creation of the University Institute of Astrophysics was announced in *BOE 3/5/75*, the Ministerial Order to create the bachelor's degree in physics at the University of La Laguna was announced in *BOE 10/3/78*, and that announcing the Programme for Studies in Astrophysics in *BOE 13/3/78*.

Chapter 8

Astrophysics in Spanish Universities (1970–1985)

Spanish university degree structure prior to the Royal Decree 1393/2007 of 29 October is described in the Wikipedia page https://es.wikipedia.org/wiki/Enseñanza_universitaria_en_España (the English version of the page offers no information prior to 2007). The bachelor's degree in physics at the University of La Laguna was created by ministerial order (*BOE 10/3/1978*). Full details of the structure of the degree are given *BOE 6/5/1978*. An overview of astrophysics and the way in which it differs from classical astronomy is given by F. Sánchez, *La Astronomía y la Astrofísica* (Acento, 1994).

Chapter 9

The Reluctant Chancellor (1976–1980)

The social and political roots of Spain's transition from dictatorship to democracy are described in S. Balfour's *Dictatorship, Workers and the City. Labour in Greater Barcelona since 1939* (Oxford: Clarendon Press, 1989). The Law for

Political Reform, passed on 18 November 1976, was published in *BOE 4 y 5/1/1977*. Concerning the expulsion of Lluis Llach from Tenerife by the police, see an interview with Ferran Monegal in *La Vanguardia* of 3 March 1976. Some of the official documentation concerning the killing of Javier Fernández Quesada has been published in the online periodical *El Observador* (www.revistaelobservador.com/79-fernandez-quesada). The Organic Law 11/1983 of University Reform was published in *BOE-A-1983-23432*.

Chapter 10

A Multinational Astrophysics Treaty (1970–2014)

For a discussion on the optical effects of occasional Saharan dust invasions, see the contributions of A. Jiménez and J. González , 'The grey behaviour of Saharan dust', and M. A. Guerrero, et al., 'Extinction over the Canarian Observatories: the limited influence of Saharan dust', in C. Muñoz-Tuñón (ed.), *Site Properties of the Canarian Observatories* (*New Astronomy Reviews*, 42, 1998) , 395–572. An historical overview of site-testing campaigns in the Canary Islands is given by F. Sánchez, 'Astronomy in the Canary Islands' (*Vistas in Astronomy*, 28), 417. Early plans for building a 60-inch infrared flux collector (which is now the Carlos Sánchez Telescope) are discussed by J. Ring, 'Infrared Astronomy, Inaugural Lecture as Professor of Physics' (Imperial College of Science and Technology, University of London, 1968), 210. A summary of the early days of astrophysics in Spain is provided by F. Sánchez and J. M. Quintana, 'Astronomy in Spain', *Anglo–Spanish Quarterly Review* (1976). In his autobiography, Fred Hoyle, *Home Is Where the Wind Blows* (Mill Valley: University Science Books, 1994), 315–20, identifies the Canaries as one of very few places on Earth where large telescopes might be usefully located and describes the long history of the Isaac Newton Telescope from its inception in 1948 to its eventual emplacement on La Palma. The first decade of science across the electromagnetic spectrum done in the Canaries after the signing of the International Agreements of Cooperation in Astrophysics is described in J. V. Wall (ed.), '10th Anniversary of the Canary Islands Observatories', *Astrophysical Letters and Communications*, 28 (2–4), 1991. For a detailed description of the 8-m telescope originally proposed for Roque de los Muchachos, see P. Álvarez Martín et al. The *ORM 8-m Telescope* (La Laguna: GRANTECAN, S.A., 1995).

Chapter 12

Science Meets Politics: Laws Governing the IAC and Its Avatars (1975–2011)

The Royal Decree-Law 7/1982 creating the IAC and defining its juridical regime ('The Astrophysical Law') was published in *BOE 5 de mayo de 1982*. Royal Decree 2678/1982, indicating the competencies of the IAC was published in *BOE 259 de 28 de octubre de 1982*. A number of laws restricting the autonomy of the IAC were subsequently passed, in particular Royal Decree 795/1989 of 23 June, concerning the organization and running of the IAC. Law 14/2000 concerning fiscal, administrative, and social order measures was published in *BOE 313 de 30/12/2000*. Law 14/2011 of 1 June 2011 (The Law of Science, Technology and Innovation) was published in *BOE-A-2011-9617*.

Chapter 13

Royalty and Heads of State Above the Clouds (1985)

Steven Weinberg's talk ('Origins') was published as a pamphlet by the IAC.

Chapter 14

The Sky Law (1978–2017)

The value of the Canarian skies as a natural resource for scientific exploitation is identified by F. Sánchez, 'El cielo de Canarias como recurso natural de reconocido interés científico', *I Jornadas de Estudios Económicos de Canarias* (Universidad de La Laguna and Banco de Bilbao, 1980). Law 31/1988 of 31 October concerning the protection and astronomical quality of the Observatories of the Instituto de Astrofísica de Canarias was publish in *BOE-A-1988-25332*. Further information on the work of the Sky Protection office is available at www.iac.es/en/observatorios -de-canarias/sky-protection (email: OTPC@iac.es).

Chapter 15

The IAC: A Dream Come True (1971–Present)

The author's address at the First Economics Studies Workshop was published in *Canarias ante el cambio* (La Laguna: Banco de Bilbao, 1981). The study of the social and economic impact of astrophysics in the Canaries referred to is by Juan José Díaz Hernández, *El impacto económico y social de la Astrofísica en Canarias* (La Laguna: IAC, 2018).

Chapter 16

Towards Excellence in Research (1971–Present)

Apart from *Prospettive dell' astronomia ottica* ('Perspectives in optical astronomy'), Atti Fond. Giorgo Ronchi, Ital.,1968, 23, n° 4, 467–74, there have been many other such 'perspectives' published, including one by the US National Research Council, *Astronomy and Astrophysics in the New Millenium, Panel* Reports (Washington, D.C.: National Research Council) and C. Warden (ed.), *Science with 8–10 m telescopes in the era of ELTs and the JWST* (La Laguna: IAC, 2009). The author describes the role of the IAC and its Observatories in 'The IAC: Its role in leading the development of Spanish astrophysics', in A. Heck, (ed.), *Organizations and Strategies in Astronomy*, vol. 5 (Dordrecht: Kluwer, 2004), 61–82, 'Astronomy in the Canary Islands', *Vistas in Astronomy*, 28 (1985), 417, and 'The Canary Islands' astrophysical observatories—a European Community project' (*Memorie della Società Astronomica Italiana*), 57 (1986), 655, and Sánchez and J. V. Wall, 'The Canarian Observatories', *Astrophysical Letters and Communications*, 28 (1991), 47. See also, F. Sánchez, 'The Instituto de Astrofísica de Canarias (IAC): its role in leading the development of Spanish astrophysics', *Astrophysics and Space Science*, 310, (2004), 61. The 'Book of Wisdom' referred to is *Recommendations about the Future of Research at the IAC* (La Laguna: IAC, 1983). G. Münch, A. Mampaso and F. Sánchez (eds), *The Universe at Large: Key Issues in Astronomy and Cosmology* (Cambridge: Cambridge University Press, 1997) is the published proceedings of the 1995 encounter funded by the BBV Foundation.

Chapter 17

Astronomical Instrumentation, Technology Transfer and Its Impact on Industry (1974–Present)

'Let others invent' is paraphrase of a popular quote from Unamuno that reflected his conviction that Spain should leave science and technology to other Europeans, while Spain concentrated on leading the world of letters. His exact words were: *Inventen pues ellos y nosotros nos aprovecharemos de sus invenciones* ('let *them* invent, then, and we'll make use of their inventions'). The discovery of Teide 1, the first confirmed brown dwarf, was published by R. Rebolo, M. R. Zapatero Osorio, and E. L. Martin, *Nature*, 377 (1995), 129. Plans for the Imperial College's 60-inch infrared flux collector are described by J. Ring, 'Infrared Astronomy, Inaugural Lecture as Professor of Physics' (Imperial College of Science and Technology, University of London, 1968), 210; B. V. Barlow, *The Astronomical Telescope* (London: Wykeham Publications, 1975), 163–4; and Tony Jones, 'El Teide and the flux collector', *J. British Astronomical Association*, 88 (1978), 257–66. A description of ISOPHOT and some of the early scientific results is provided by M. F. Kessler, 'The first results from ISO', *ESA Bulletin* 86 (1996). A description of EMIR is given in the project's webpage: https://www.iac.es/en/observatorios-de-canarias/telescopes-and-experiments/espectrografo-multiobjeto-infrarojo. The OSIRIS instrument is described in the webpage https://www.iac.es/en/observatorios-de-canarias/telescopes-and-experiments/osiris-nasmyth-b. The COSMOSOMAS experiment is described by J. E. Gallegos, et al., 'COSMOSOMAS: a circular scanning instrument to map the sky at centimetric wavelengths', *Monthly Notices of the Royal Astronomical Society*, 327, 1178–86 (2001). A description of the QUIJOTE experiment is provided by J. A. Rubiño-Martín, 'The QUIJOTE experiment: project status and first scientific results', in S. Arribas et al., *Highlights on Spanish Astrophysics IX, Proceedings of the XII Scientific Meeting of the Spanish Astronomical Society* (London: EDP Sciences, 2016), 99–107. The activities of the Technology Transfer Office (OTRI) are summarized in the IAC's annual *Memoria*. The economic impact of large telescopes is described by J. Burgos-Martín, M. Sánchez-Padrón, F. Sánchez, and C. Martínez-Roger, 'Extremely large telescopes as a motor of socio–economic development and implications of its construction and installation', SPIE Proceedings 5382: 2nd Bäckaskog Workshop on Extremely Large Telescopes, 142 (2004).

Chapter 18

The Rise of Space Astrophysics in Spain (1942–Present)

The efforts of Spanish engineers to involve scientists in their space programmes during the Franco era is described by Lino Caprubí, *Engineers and the Making of the Franco Regime* (Cambridge, Mass.: MIT Press, 2014). A brief overview of Spain's space activities is given in the doctoral thesis of Iván Fernández Pérez, *Aproximación histórica al desarrollo de la astronomía en España* (University of Santiago de Compostela, 2009). More detailed treatment is provided by J. M. Dorado, M. Bautista, and P. Sanz Aránguez, *Spain in Space: A Short History of Spanish Activity in the Space Sector* (Nordwijk: ESA, 2002); F. González, 'La entrada de España en la Era Espacial: La instalación de la
estación de seguimiento de satélites en el Observatorio de San Fernando (1956–1958)', in M. Vallejo (ed.), *Jornadas Científicas 250 años de Astronomía en España* (San Fernando: Real Instituto y Observatorio de la Armada, 2003); J. M. Sánchez Ron, 'Terradas y la aeronáutica', *Revista Quark*, 31, (2004), 71–7; and J. M. Sánchez Ron, *50 años de ciencia y técnica aeroespacial* (Aranjuez: Ministerio de Defensa, 1997).

Chapter 20

Bringing Astronomy to the Public (1986–Present)

Ample details of the IAC's extensive outreach programme are given on the Institute's webpage: www.iac.es.

Chapter 21

The Biggest, Most Advanced Telescope in the World (1989–Present)

A general overview of the GTC is given by F. Sánchez, 'Gran Telescopio Canarias' in *Future Large Scale Facilities in Astronomy*, 23rd meeting of the IAU, Joint Discussion 9 1(997), A full description of every aspect of the design of the GTC, together with a summary of the site selection campaign, is given in P. Álvarez Martín et al., *Gran Telescopio CANARIAS: Conceptual*

Design (La Laguna: GRANTECAN, S.A., 1997). The first scientific results of the telescope were presented by J. M. Rodríguez Espinosa, F. Garzón López and V. Melo Martín (eds), *Science with the GTC 10-m Telescope* (*Revista Mexicana de Astronomía y Astrofísica Serie de Conferencias*, 16, 2003). Asteroid 44103 (Aldana) is listed in L. D. Schmadel's *Dictionary of Minor Planet Names: Addendum to Fifth Edition 2003–2005* (Berlin: Springer Verlag, 2006), 209–210.

Chapter 22

The European Extremely Large Telescope: Dealings with ESO (1961–Present)

An early discussion of seeing is given by J. Stock and G. Keller in G. P. Kuiper (ed.), *Telescopes* (Chicago: Chicago University Press, 1960), and site selection is covered in the same volume by A. B. Meinel. A description of the 100-m OWL is provided in the *OWL Concept Design Report: Phase A Design Review* (Garching: ESO), The Euro-50 telescope proposal is described by T. Andersen, A. Ardeberg, and M. Ower-Petersen (eds), *Euro 50: Design Study of a 50 m Adaptive Optics Telescope* (Lund: Lund University, 2003), and the E-ELT is described in *The E-ELT Construction Proposal* (Garching: ESO). The European Northern Observatory concept is described by D. Leverington, *Observatories and Telescopes of Modern Times: Ground-Based Optical and Radio Astronomy Facilities since 1945* (Cambridge: Cambridge University Press, 2016), ch. 6. See Muñoz-Tuñón (ed.), *Site Properties of the Canarian Observatories* (New Astronomy Reviews, 42,1998) for a full discussion of site conditions at Roque de los Muchachos Observatory. A summary of dust episodes at Mauna Kea is provided by M. Dickinson, C. Dumas, and M. Bolte, *NOAO Newsletter* 115 (March 2017), 14–16. A comparative site characterization study of Armazones and Paranal for the emplacement of the VLT is given in M. Sarazin (ed.), VLT Report No. 62, 'VLT Site Selection Working Group Final Report' (Munich: ESO, 1990). There is no similar report for the E-ELT. A discussion of site characterization efforts worldwide is provided by Vernin, Benkhaldoun, and Muñoz-Tuñón (eds), *IAU Technical Workshop: Astronomical Site Evaluation in the Visible and Radio Range* (San Francisco: ASP, 2002).

Chapter 23

The Starlight Foundation: A Step Beyond Astrotourism (2008–Present)

The author's contribution to the 1995 conference on sustainable tourism was entitled 'La astronomía, experiencia en marcha de un turismo cultural en Canarias', in *Sustainable Tourism World Conference* 1995, 77. The 2007 Starlight Conference proceedings were edited by J. Jafari and C. Marín, *Starlight Conference 2007*. The Declaration in Defence of the Night Sky and the Right to Starlight may be downloaded from: https://cds.cern.ch/record/1481596/files/978-1-4614-3822-9_BookBackMatter.pdf. Concerning Commander J. Cousteau and the Unesco meeting see K. MacFarlane, 'Los derechos humanos de las generaciones futuras (la contribución jurídica de J. Cousteau)', *Ultima Década*, 8, 145–65. The work of the Starlight Foundation is fully described in its website: https://fundacionstarlight.org/.

Chapter 24

Astrophysics in Spain: The Wider Picture (1974–Present)

A general overview of modern astronomy in Spain is provided by F. Sánchez and P. Murdin, 'Spanish astronomy', in P. Murdin (ed.), *Encyclopedia of Astronomy and Astrophysics* (London: Institute of Physics, 2001). Iván Fernández Pérez, *Aproximación histórica al desarrollo de la astronomía en España* (University of Santiago de Compostela, 2009), ch. 3 and 4, gives a detailed description of many of the observatories and research centres and university departments that have arisen in Spain over the past half century. Enrique García Gómez describes the history of astronomy in Toledo in *Diez Siglos de Ciencia y Científicos Toledanos* (Ediciones Covarrubias, 2018). The importance of EU funding for scientific cooperation among member states is covered by F. Sánchez, 'European Union funding stimulates cooperation among European astronomers', *European Astronomical Society Newsletter*, 9 (1994). In celebration of the 2009 International Year of Astronomy the *Revista Española de Física* of the Spanish Royal Society of Physics dedicated volume 23 of that publication to mark the event under the title 'Año International de la Astronomía'.

Index

A

Aguirre, E., 218
Airglow, 42–43
Alberti, R., 197
Aldana, F., 219, 220, 233
Alexander, J., 82
Alfonso the Wise Council (CSIC), 139
Ali, 241
Almeida, María ('Mary') Anselma, 8, 10, *20*, 48
Álvarez, P., 164, 213, 216, 227–228
American Dark Sky Association (DSA), 253
Ángeles, V. de los, 126
Ångström, K.
    solar constant measurements in Tenerife (1895-96), 26
Aragon Centre of Cosmic Physical Studies (CEFCA), 267
Arango, P., 200
ARENA, 177
Ariño, G., 106
Armas, A. de, 112
Armstrong, N., *210*
Arnay, R.
    Commander of the Order of Civil Merit, 123

Ascanio Togores, R., 227
Assessorial Committee for Large Scientific Facilities
    scepticism over GTC project, 218
Association of Universities for Research in Astronomy (AURA), 149
Astrobiology Centre (CAB), 267
Astronomical site testing, 29, 34, 45, 50, 53, 82, 86, 232, 242
Astrophysics internships, *67*
Astrophysics Law, 106–108, 142, 152
Astrotourism, 254
Atacama Large Millimeter/ submillimeter Array (ALMA), 229
Atmospheric sounding rockets, 179
    FOCCA mini-photometers, 181, 185
Atmospheric transparency
    measurement of, 39, 46, *47*, 79
Atmospheric turbulence, 2
    increase with primary aperture, 212
Automatic multichannel telescope, 163
Autonomous University of Madrid, 268
Aznar, J.M., 219

B

Baade, W., 82
Bachelet, M., 243
Barbier, D., 34, 40
Barcons, X., 232, 240, 244–246
Barreda, F. de la, 219
Battaner, E., 68, 180, 181, 263
Beckwith, S., 154
Bethencourt, A., 72, 75, 76, 100, 170
Big Science, 129, 196, 208, 213–216,
        222, 224, 269
Birt, W.R.
    lunar nomenclature, 22
Boksenberg, A., 213
Bonet, J.A.
    Commander of the Order of Civil
        Merit, 123
Borda, J.C.
    astronomical observations in
        Tenerife (1778), 22
Bordeaux Telescope, 38–43, 80, 151
BOSS (SDSS-III), 186
Boyer, B., 128
Brück, H., 231
Burbidge, G., 156
Burbidge, M., 154–156
Busquin, P., 231

C

Cabrera, N., 142
CAI, 153, 154
CAIN, 166
Calar Alto Observatory, 109, 140,
        260, *261*, 262
Caltech, 149
Canada-France-Hawaii Telescope, 81
Canal, R., 263
Canarian independence movements, 76
Canarian Observatories, xi, 38, 95,
        106, 109, 119, 129, 132, 134,
        142, 144, 152, 177, 196, 232,
        233, 242, 248, 270
    generators of new technology, 162

open days and guided tours, 195
Canarian skies as a natural resource,
        85–88, 120, 140, 142, 162,
        209, 261
Canary Islands
    technological and industrial
        development, 143
Canary Islands Winter School of
        Astrophysics, *189, 190,
        192, 193*
    rationale, 190
Caprubí, L., 180
Carlos Sánchez Telescope, 81
Carlsberg Meridian Circle, 88
Carrasco, P., 55
Carrillo Kábanas, A., 131
Castro, J., 213
Cavendish Laboratory, 84, 168–169
Central Democratic Union, 72
Centre of Energetic, Environmental,
        and Technological
        Research, 267
Cepa, J., 166
Cerro Armazones, 240–242
    discounted as site for TMT, 241
Cerro Honar, 241
Cerro Vicuña, 241
Cesarsky, C., 233–236
CHEOPS, 187
Cherenkov Telescope Array (CTA), 80,
        149, 232
    CTA North, 232
    CTA South, 232
Chinese Space Agency, 186
Chirino, M., 120
Cid Palacios, R., 58, 65, 139, 265
Claret de Fleurieu, C.P.
    astronomical observations in
        Tenerife (1769), 22
Classical astronomy, 57, 63, 137, 264
Clavijo, R., 141, 146
Cloud cover
    measurement of, 45
CMB anisotropies, 168

Cold War
    strategic importance of Spain for US
        interests, 180
Collados, M., 162
Colonialism in science, 87, 140, 234
Comet Hyakutake
    nationwide contest organized by
        IAC, 205
    public viewing from Puertito de
        Güímar, 204
Comet Schumaker-Levy
    first report of collision with Jupiter
        from Teide Observatory, 204
    Public broadcast of collision from
        Science and Cosmos
        Museum, 204
Communist Party legalized, 72, 73
Complutense University of Madrid, 50,
        62, 127, 163, 268
CONIE, *see* National Commission of
        Space Research
CONSTELLATION, 177
Convent of San Francisco
    restored as part of IAC
        inaugurations, 127
CoRoT, 187
Corradi, R., 227
Cosmic Background Explorer, 168
Cosmic Microwave Background
        (CMB), 167
COSMOlogical Structures On
        Medium Angular Scales, *see*
        COSMOSOMAS
COSMOSOMAS, 168
Costillo, L., 181
Coup d'état attempt of 23 February
        1981, 107
Cousteau, J., *252*
CSIC, 5, 6, 58, 86, 105–114, 139,
        141, 214, 218, 244, 262, 265
    centralizing tendencies, 105
V404 Cygni
    IAC confirmation of black hole
        component, *157*

Daniel K. Inouye Solar Telescope, 162
Danish royalty, 119
Darwin, C.
    quarantined off Tenerife (1832), 22
Dawkins, R., *210*
Daza de Valdés Optics Institute, 5, 161
Declaration in Defence of the Night
        Sky and the Right to
        Starlight, 250
Department of Communications
        Engineering, 169
de Zeeuw, T., 232, 241
Díaz, J.J., 143
Diffraction rings around stars
    a measure of atmospheric
        turbulence, 46
Dommanget, J., 29, 34, 230
Duke and Duchess of Gloucester, 119
Dumont, R., 38, *40*, 43
Duncan Piazzi Smyth, J., 24, *25*
Duperier, A., 55
Dutch royalty, 119

e-BOSS (SDSS-IV), 186
Ebro Observatory, 56
Echelle SPectrograph for Rocky
        Exoplanets and Stable
        Spectroscopic Observations
        (ESPRESSO), 187
1959 eclipse of the sun, 1–5, 28
    international expeditions, 1, 5
El Astrofísico, see Instituto de Astrofísica
        de Canarias
El Mercurio, 243
El Paso abstract movement, 120
ESA, *see* European Space Agency (ESA)
Espectroflal
    awarded Agustín de Bethencourt
        Prize for Research, 170
Esteban Terradas National Institute of
        Aerospace Technology, 265

ESTEC, 184
EUCLID mission
  NISP instrument, *187*
EURO-50, 236
European Astronomical Society, 266
European Extremely Large Telescope,
    80, 177, 219, 226, 229–246
  apathy of Spanish government over
    E.ELT site, 243
  IAC astrophysicists barred from
    E-ELT emplacement
    negotiations, 231
  lower costs if located in La
    Palma, 239
  proactive stance of Chilean
    government to bring it to
    Chile, 243
European Northern Observatory,
    231, 235
  viewed with suspicion by ESO, 232
European Observatories
  incorporating ESO and ENO, 232
European Solar Telescope, 162
European Southern Observatory, 29,
    149, 229
  Atacama Desert, 229
  La Silla, 229
  Paranal, 229
  Spain's entry, 229
European Space Agency (ESA),
    180–187, 232, 244, 265
European Space Research Organisation
    (ESRO), 139, 180
European VI Framework
    Programme, 240
Exozodiacal light, 44
  discovery with VLTI, 38
Extragalactic light, 43

F

Faculties of Physics and
    Mathematics, 141
Fernández Caldas, E., 51, 141

Fernández Quesada, J.
  killed on ULL campus, 74
Fernández, M.J., 263
Fernández, P., 219
Feuillée, L.
  Tenerife astronomical observations
    (1724), 22
FIN, 166
First National Assembly of Astronomy
    and Astrophysics, 265, *266*
First Spanish university course in
    astrophysics, 63
Flux Collector, *see* Carlos Sánchez
    Telescope
Fraile, M., 97
Francisco, J.C., 219
Franco regime
  hindrance to international
    cooperation, 79, 81
Fraunhofer Institute, 83, 151
Fröhlich, C., 184

G

Galán, M., 146, 164, 181
  Commander of the Order of Civil
    Merit, 123
Galeán, J.M., 89
García-Cuenca, T., 219
García de la Rosa, I., 198–200
Garmendia, C., 115, 218,
    224, 244–245
Garzón, F., 166
  ISOPHOT-S, 182
Gegenschein, 43
Geographical and Land Registry
    Institute, 56
Giacconi, R., 233
Gilmore, G., 154
Glover, J., 224
González, F., 111
González, J.M., 214
GRANTECAN, S.A., 216,
    219–221, 227

Gran Telescopio Canarias, xi, 42, 129,
          152, 162, 164–167, 172, 175,
          177, 206, *210*, 212–227, 230,
          231, 234, 238, 250, 267, 269
     backing of IAC's Governing Council
          required, 214
     benefits for Spanish high-tech
          industry, 212
     British approached for
          collaboration, 213
     British opt for Gemini project, 214
     change to 10-m segmented primary
          design, 217
     FEDER funds feasibility study, 214
     first light, 222
     initial plan for 8-m monolithic
          primary, 216
     invitation for international
          tenders, 218
     missed opportunity by Spanish firms
          to build segmented
          primary, 212
     no budget overrun, 224
     optimization for visible and near
          infrared, 217
     participation of national firms in
          construction of GTC, 172
     role in developing high-tech
          industry in Spain, 162
     running costs
          underfunded, 225–226

Halley's Comet
     return of 1986, 197
     star party on Las Teresitas beach,
          197, *198*
Hanle, 241
Hannover Universal Exhibition
     La Palma exhibit, 201
Haute-Provence Observatory,
          34, 38–40
Hawking, S., *160*

HELAS, 177
Helioseimology, 183
Hendaye
     Spanish migrant workers, 30
Hermoso, Manuel, 214
Herschel, 177, 185–187
Hewish, A., 84
Hewlett Packard
     campaign of denigration against
          NOVA, 163
Hoyle, F., 82

IAC Noticias magazine, 203
IAC Observatories, 65–67, 80–81,
          106, 231, 236, 237, 248, 251
IAC-80 Telescope, 164, 165, 213
ICONA, 89, 90, 98, 203
Imax, 177
Imperial College of Science and
          Technology, 81, 166, 182
Inaugurations of the IAC and its
          Observatories, 144
     security concerns, 121
Industry, External Trade, Research, and
          Energy Committee, 235
Infrared astronomy, 24, 41, 81,
          166, 182
Infrared Space Observatory, 166, 182
     launched in 1995, 183
Institute of High Energy Physics, 267
Institute of Physics of Cantabria, 185
Institute of Space Sciences, 267
Instituto de Astrofísica de Andalucía,
          181, 262–267
Instituto de Astrofísica de Canarias, xi,
          59, 67, 71, 72, 76, 102,
          105–116, 119, 122, 169,
          173, 209, 254, 265,
          267, 269
     accreditation as a Severo Ochoa
          Centre of Excellence, 159
     burden of excessive bureaucracy, 148

Instituto de Astrofísica de
    Canarias (*cont.*)
  capacity to hire research staff
    removed, 113
  Computer Centre, 161, 163, 164
  Director's Support Team, 148, 196,
    204, 247
  External Assessment
    Committee, 153
  extra-budgetary financial
    support, 148
  first headquarters, 145
  Governing Council, 109, 144, 155,
    157, 175
  inauguration of
    headquarters, 122–125
  Instrumentation Division, 42, 162,
    164, 213, 227
  juridical personality, 105, 108
  laws restricting IAC's capacity to
    act, 115
  legal status as a consortium, 111,
    115, 145
  Management Committee, 141
  mission, 108
  need for a foundation to
    complement outreach
    activities, 247
  net contributor to Spain's GDP, 260
  new headquarters, 146
  outreach overload, 247
  Research Division, 157
  scientific editorial service, 157
Instituto Universitario de
    Astrofísica, 145
Institut Royal Météorologique, 184
INTA, *see* National Institute of
    Aeronautic Technology
Integrated starlight, 43
Intelligent Specialization Strategy
    (RIS3), 149
Inter-island Association of Cabildos of
    the Province of Santa
    Cruz, 139

International Agreements concerning
    Astrophysics, *see* Treaty
    concerning Cooperation in
    Matters of Astrophysics
International Astronomical Union
  lunar nomenclature, 21
  support for Sky Law, 132
International Commission on
    Illumination
  support for Sky Law, 132
International Scientific Committee
    (CCI), 88, 121, 132,
    203, 213
  recriminations over use of name
    ENO, 236
Interplanetary dust particles
  Doppler shifts of, 40
Interplanetary medium, 37, 38,
    43, 62, 65
Interstellar medium
  enrichment by stellar ejecta, 35
Isaac Newton Telescope, 82, 86
Island Cabildo of Tenerife, 145, 162
ISOPHOT-C, 182
ISOPHOT-P, 182
ISOPHOT-S, 166, 182–183, *184*, 185
  project shelved on death of Carlos
    Sánchez, 182
  resumed under supervision of
    F. Garzón, 183
Israelian, G., *210*
Iye, M., 217
Izaña Meteorological Observatory, 5,
    16, 26, 38, 46, 80, 180
  staff, 19
Izeta, I., 9, 11

J

Japanese Space Agency, 182
Jodrell Bank Observatory, 168–169
Joint Organization for Solar
    Observations (JOSO), 82
Juan March Foundation, 163

K

Kaiser's Lodge, 10, *11*, 13, 26
Keck telescopes, 212, 221
Kiepenheuer Institute, 81
Kiepenheuer, K.O.
    aborted solar observatory on Mt
        Teide, 84
King Juan Carlos I, 86, 123, 124, 224
Kron, E.
    denied access to Tenerife
        summits, 27

L

Laboratory of Natural Sciences
    now the Prado Museum, 55
Labour lawyers murdered in
    Madrid, 74
Langrenus (M.F. van Langren)
    lunar map, 21
Lapesa, R., 200
Large Binocular Telescope, 224
Law concerning Science, Technology,
    and Innovation, 115
Law for the Protection of the
        Astronomical Quality of the
        Observatories of the IAC,
    see Sky Law
Law of Astrophysics, 147, 190
Law of Political Reform (1976), 71, 86
Law of Science, 111–113
    limits IAC's capacity to act, 113
Leonov, A., *210*
Light pollution, 38, 42, 83, 86, 131,
    132, 133, 253–255
    LED light fixtures and their hazards,
        135, 251
    LED lighting, 135
Limiting magnitude
    as measure of sky transparency, 46
Liverpool John Moores University, 206
Llach, L.
    banned concert, 72
Longair, M., 156

López, J.J., 181
Los Rodeos air crash, 74
Lunik 2, 2
Lynden-Bell, D., 156

M

Mack, B., 213
Madrid Institute of Material
    Sciences, 267
Manrique, C., 120, 125
    controversy over observatory
        monument in La Palma, 121
Maravall, J.M., 111–113
    attempt to undermine IAC's
        juridical status
        caricatured, *112*
Mardones, L., 112
Mare Imbrium, 21
Martín, A., 197, 200
Martín Arias, J., 111
Martínez, C., 158, 164
Martínez Esteruelas, C., 262
Martínez Sáez, L., 196, 203, 249, 258
Martínez, V., 162
Mascart, J.M.
    Tenerife expedition (1910), *11*,
        14, 26–27
Mauna Kea Observatory, 82
Max Planck Institute for
        Astronomy, 182
Max Planck Institutes, 149
May, B., 40, 41, *210*, 223
May 1968 Paris disturbances, 32
Mayor Zaragoza, Federico, 141
Mendoza, J., 219
Menéndez y Pelayo University,
        142, 192
Micrometeoroids, 43
Minor Planet 44103 (Aldana), *220*
Modern astronomy, 2, 57, 87, 137,
        261–263, 268
Moles, M., 263
Mons Telescope, 81

Moscoso, J., 111, 112
Mountain, M., 217
Müller, G.
  denied access to Tenerife
    summits, 27
Multi-object Infrared Spectrograph
  (EMIR), 166–167
Münch, G., 154–156, 216
Muñoz Tuñón, C., 216, 237, 241

N

Narro, J., 224
NASA, 149, 179–187
  tracking stations in Spain, 179
National Astronomical
  Observatory, 267
National Commission of Astronomy,
  109, 139, 154, 259–265
National Commission of Space
  Research (CONIE),
  163, 180–181
National Geographical Institute,
  109, 265
National Institute of Aeronautic
  Technology (INTA),
  179–181, 185
National Institute of
  Astronomy, 58, 139
National Institutes of Natural Sciences
  of Japan, 229
National Observatory, see Astronomical
  Observatory of Madrid
National Observatory of
  Madrid, 56, 138
National Plan for R&D, 154
National plan for the training of
  astrophysicists, 61, 262
National Programme for Astronomy
  and Astrophysics, 244, 267
National University of Distance
  Education, 268
Nelson, J., 217

Nocturnal Sky Section, 41
Nombela, C., 218
Nordic Optical Telescope, 204
Northern Hemisphere Observatory
  (NHO), 82, 86
NOVA 1200, 161
Novikov, I., 156
Nuclear Energy Council, 7, 46, 163

O

OAS (Organisation de l'Armée
  Secrète), 32
Ochoa, S., 200
  Nobel Prize for Medicine, 2
Ophthalmology Service of the
  University Hospital, 172
OSIRIS, see Optical System for
  Imaging low Resolution
  Integrated Spectroscopy
  (OSIRIS)
Optical turbulence, 45, 79
Optical System for Imaging low
  Resolution Integrated
  Spectroscopy (OSIRIS), 165,
  166, 223
Oriol, I. de, 200
Orús, J.J. de, 58, 138, 139, 263–265
O'Shea, P., 200
Osterbrock, D., 156
Other universities in Madrid, 268
Outreach, xi, 41, 90, 107, 134, 144,
  156, 247, 248, 255, 269
  budgetary and bureaucratic
    constraints, 207
  Canarias innova radio
    programme, 206
  Caosyciencia digital magazine, 206
  combatting pseudoscience, 195
  GTCdigital magazine, 206
  IAC profile in social media, 207
  material for amateur associations
    and science popularizers, 206

repaying the taxpayer, 195
robotic telescopes for schools, 206
OverWhelmingly Large telescope
    (OWL), 234–236

P

Pagel, B., 156
Paris Institute of Astrophysics, 29
Pérez Llorca, J.P., 91, 92, 143
Physics Faculty (ULL), 62, 72
Physics Institute of Cantabria, 169, 267
Piazzi Smyth, C.
    Tenerife expedition (1856),
        22–26, *27*, 38
Pico, 21, 22, 83
Pico Teneriffe, *see* Teide, Mt
Piñera, S., 243
Pingré, A.G.
    astronomical observations in
        Tenerife (1778), 22
Planck, 149, 169, 177, 185–187
    Radiometer Electronic Box
        Assembly, 186
Plato, *186*, 187
Polanco, Jesús de, 200
Political unrest during Transition, 72
Polytechnic University of
    Cartagena, 268
Polytechnic University of
    Catalonia, 268
Polytechnic University of
    Valencia, 268
Postgraduate Teaching Division
    (IAC), 69
Presidency of the Government of
    Spain, 146
President of the Federal Republic of
    Germany, 119
President of the Republic of
    Ireland, 119
Press and Outreach Unit, xv, 196, 203
    training of science journalists, 204

Primo Yúfera, E., 86, 141, 262
Prince of Asturias, 122, 197,
    219–223, 238
    made Honorary Astrophysicist of
        the IAC, 122
Prince of Asturias Awards, 101
Provincial Inter-island Association of
    Cabildos of the Province of
    Santa Cruz de Tenerife, 142
Puerto, Carmen del, xv, 203

Q

Queen Sofía, 200
Q-U-I JOint TEnerife CMB, *see*
    QUIJOTE-CMB
QUIJOTE-CMB, 169–170
Quintana, J.M., 262–263

R

Rebolo, R., 116, 168
*Recommendations about the Future of
    Research at the IAC*, 153
Redman, R.O., 7
    unpublished report, 45
Rees, M., 154–156, 214
Rego, M., 263
REOSC, 163
Research Assessment
    Commission, *see* CAI
Riccioli, G.B.
    lunar map, 21
Rivero, P., 224
Robles, E., 219
Robotic telescopes
    GLORIA worldwide array, 207
    PETeR, 206
Roca, T., 81, 113, 183–185
Rodrigo, R., 244
Rodríguez Espinosa, J.M., 216, 227
Rojo, J., 113, 114
Rollán, Á., 263

Romañá, Father Antonio, 5, 58, 139, 265
Roque de los Muchachos
   military authorities decide not to build radar station, 101
   negotiations with military authorities, 97
   prospective site for military radar station, 90, 97, 143
Roque de los Muchachos Observatory, xi, 86, 90, 101, 102, *103*, 105, 121, 125–127, *136*, 140, 143, 152, 165, 175, 191, 196, 203, 204, 206, 212, *223*, 224, 231–235, *237*, 241, 243, 248–254, 260
   building of access road, 89
   chosen site for Isaac Newton Telescope, 82
   failure of visitors' centre project, 203
   false claims of dangerous waste emissions, 248
   IAC allocates funds for building observatory, 90
   inauguration, 125–126
   international conference in defense of the Observatories, 249
   science cultural park, 196
   selected as alternative site for the TMT, 241
Roscosmos, 186
Rosenberg, A., 204
Royal Belgian Observatory, 34
Royal Convent of the Immaculate Conception
   formerly the Convent of San Francisco, 127
Royal Greenwich Observatory, 82–85, 213
Royal Naval Observatory, San Fernando, 45, 56, 88, 179, 265, 267
Royal Observatory, Edinburgh, 86
Royal Observatory of Madrid, 262–264
Royal Observatory of the Cape of Good Hope, *see* South African Astronomical Observatory
Rubbio, C., 208
Rubio García-Mina, J., *52*
Ruiz, J., 115

S

Saavedra, J., 112, 218
Sage, L., *210*
Saharan dust
   as a site-testing parameter, 79, 80, 242
   Heidelberg Observatory decides against Tenerife, 80
   hindrance to international cooperation, 79
Salazar Palace, 128, 224
   restored for IAC inaugurations, 128
Sánchez Asiaín, J.Á., 173
Sánchez, C., 41, 81, 123, 166, 182, 263
   Commander of the Order of Civil Merit, 123
Sánchez, F., *210*
   Acting Chancellor of ULL, 75–77
   Commander in Chief of the Order of Civil Merit, 123, 124
   Commander of the Swedish Order of the Pole Star, 126
   CSIC campaign to remove him as IAC Director, 114
   elected member of the IAU, 139
   entry into research, 6
   first Spanish chair of astrophysics, 138
   first visit to La Palma (1973), 83
   Founding Director of the IAC, 116
   graduation in Physical Sciences, 6
   leaves research to direct the IAC, 144

PhD thesis, 49
refusal to be appointed ULL
    Chancellor, 76
retirement, *116*, *117*
unpopular at ESO, 234
Vice-chancellor for Research, 72
visit to UK astronomical centres, 84
Sánchez, J., 83, 168
seconded to Jodrell Bank
    Observatory, 168
Sandage, A., 156
San Pedro Mártir, 241
Schröter, J.H.
lunar nomenclature, 22
Schwarzschild, K.
denied access to Tenerife summits, 27
Science and Technology Office
    (OCYT), 219–220, 233
Science education
    IAC projects for school pupils and
    teacher training, 205
Science Walk in Santa Cruz de la
    Palma, 254
Scientific Research Committee, *94*
Seara, L.G., 76
Seeing, x, 242
Selby, M., 182
Semi-automatic double telephotometer
    the 'coffee pot, 42
Service d'Astrophysique/Centre
    d'Énergie Atomique, 184
Seville Universal Exhibition, 197–201
    IAC exhibits, 201
    problems with IAC radiotelescope
    insignia, 201
Simony, O.
    variations in solar spectrum with
    height, 24
Site-testing campaigns, 50, 79–86, 230
Sky Law, 132–136, 243, 248, 251
    control of light pollution, 131
    further supportive legislation, 135
    opposing interests in La Palma, 248

prohibition of air traffic over
    Observatories, 134
prohibition of flights over Canarian
    Observatories, 131
restrictions on radioelectric
    emissions, 131
Sky transparency, 45, 46
Smoot, G., *210*
SN 1006
    identification of white dwarf binary
    as progenitor, *157*
SOLAIRE, 177
Solana, J., 142
    cultural content of the IAC
    inaugurations, 127
Solar and Heliospheric Observatory,
    183, *184*
    GOLF, 183–184
    VIRGO, 184
Solar physics, 41, 81, 168
Soulié, G., 38
Sounding rockets
    IAC/IAA payloads, *181*
South African Astronomical
    Observatory, 261
Spanish Astronomical Society, 57, 69,
    216, 259, 266, 267, 278
Spanish astrophysics
    high rate of publication, 260
    world ranking, 259
Spanish doctoral theses in
    astrophysics, 39
Spanish migrant workers, *31*
Spanish monarchs, 122
Spectrophotopolarimetric
    telescope, 39, 164
Starlight Foundation
    Corporación 5, 254, 258
    shoestring budget, 258
    Starlight Astronomical
    Monitors, *256*
    Starlight Auditors, 256
    Starlight Certifications, 253

Starlight Foundation (*cont.*)
   Starlight Monitor, *257*
   Starlight Reserves, 255
   Starlight Touristic Destinations,
      255, 257
Starlight Initiative, 253
STARMUS, 160, *210*
Suárez, A., 72, 86, 90, 99–101, 107
Sunyaev, R., 156
Swedish royalty, 119, 126
Szostak, J., *210*

T

Tarenghi, M., 217
Tarradellas, J., 31
Tarter, J., *210*
Technical Office for the Protection of
      Sky Quality, *see* OTPC
Technical Services Department
   forerunner of IAC's Instrumentation
      Division, 162
Technology transfer, 147
   ASTROTECNIA, 175
   Banco de Bilbao, 173
   Caja Canarias, 173
   Canarian Industrial
      Association, 171
   economic returns for IAC, 172
   Energía Solar Española S.A., 172
   Espectroftal, 170
   Galileo Engineering and
      Services, S.A.
      failed example of technology
         transfer, 173
   Hidrola, 174
   Hipocampus, 171
   IBERDROLA, 173–174
   Iberduero, 173, 174
   IDOM, 169
   Inasmet-Tecnalia, 175
   joint ventures with industrial
      firms, 172

MEL, 171
Oftacron, 171, *173*
outsourcing to local firms, 172
TRAGSA, 174
Teide 1
   discovery with IAC-80, 164
Teide National Park, 145
Teide Observatory, 5–7, 18, *38*, 41, *42*,
      45, *49*, 50, *52*, 58, 62, 66,
      80–86, 102, *103*, 105,
      122–123, 145, 164–170, 182,
      183, 191, 196, 202, 204, 207,
      260, 262
   re-inauguration in 1985, 122
   visitors' centre, 196
Tenerife Experiment, 168–178
Tenerife Museum of Science and the
      Cosmos, 197, *199*
   management and staffing, 200
   visit by Prince of Asturias, 197
Teneriffe Mountains, 21, 22
Texcan, 172
Thirty Meter Telescope, 241, *242*
   controversy over Mauna
      Kea, 241
Ting, S., 254
Torre, Pilar de la, 111
Torroja, J.M., 5, 6, 58, 139
Transition to Democracy, 71
Treaty concerning Cooperation in
      Matters of Astrophysics, 72,
      92, 101, 131, 141, 143,
      146, 190
   an attempted renegotiation of the
      Treaty, 94
   celebrating successful conclusion of
      Level 3 negotiations, *91*
   detailed summary, 90
   personnel needs of IAC, 108
   scope of agreement, 92
   signing of Treaty, 91, 92
Trujillo, G., 75
   elected Chancellor of ULL, 77

U

ULL–IAC Postgraduate School, 67
Unamuno, M. de, 55, 161, 170
UNESCO, 190, 250–255
United Nations Universal Declaration
    of the Human Rights of
    Future Generations, 250
Universal Declaration concerning the
    Rights of Future
    Generations, *252*
*The Universe at Large (Key Issues in
    Astronomy and Cosmology)*,
    156, 158
University Institute of Astrophysics, 47,
    51, 61, 62, 81, 105, 108,
    139, *140*, 141, 162, 195
University Miguel Hernández, 268
University of Alcalá de Henares, 268
University of Alicante, 268
University of Barcelona, 267
University of Bordeaux, 38, 80, 184
University of Cádiz, 268
University of California, 149
University of Cambridge,
    149, 168–169
University of Cantabria, 268
University of Extremadura, 268
University of Girona, 268
University of Granada, 68, 185, 268
University of Huelva, 268
University of Jaen, 268
University of La Coruña, 268
University of La Laguna, 5–7, 50, 51,
    61–69, 72–77, 82, 85, 100,
    105–109, 113, 122, 139,
    141–145, 161, 162, 170–172,
    176, 214, 218, 260–268
    Commune, 74, 75
    Department of Economics,
        Accountancy, and
        Finance, 143
    Faculties of Physics and
        Mathematics, 64

Faculty of Medicine, 171
first Spanish university to elect its
    chancellor democratically, 77
University of La Rioja, 268
University of Manchester,
    149, 168–169
University of Mons, 81
University of Murcia, 268
University of Nice, 184
University of Oviedo, 268
University of Salamanca, *267*, 268
University of Santiago de
    Compostela, 268
University of the Balearic
    Islands, 268
University of the Basque
    Country, 268
University of Valencia, 268
University of Valladolid, 185, 268
University of Vigo, 268
University of Zaragoza, 65, 268
University Rovira i Virgili, 268
Upper Council for Science and
    Research, *see* CSIC
US National Science Foundation
    (NSF), 229

V

Vacuum Tower Telescope, 81, 165
Varela, T., 237, 255
Vázquez, M., 81, 123
    Commander of the Order of Civil
        Merit, 123
Verdun de la Crenne, J.R.
    astronomical observations in
        Tenerife (1778), 22
Vernet, J., 128
Vernin, J., 241
Very Large Telescope (VLT), 177,
    217, 225
Very Small Array, 169
Vidal, S., 181

W

Weinberg, S., 122
Westendorp, C., 235
William Herschel Telescope, 165,
    204, 213
World Radiation Centre, 184
World Tourism Organization, 255

Z

Zalote, M., 9–12, 47, 49
Zapatero, J.L., 115, 218, 240, 243, 244

ZARA-INDITEX, 260
Zodiacal cloud, 37–44
Zodiacal dust
    interaction with solar
        plasma, 44
    Poynting–Robertson
        drag, 44
    Poynting vector, 44
Zodiacal light, 19, 23, 37–44,
    81, 163
    permanence, 44
Zwaan, C., 154